PRACTICAL GOAT-KEEPING

JOHN & JILL HALLIDAY

WARD LOCK LIMITED · LONDON

Acknowledgements

The authors thank Kaye and Johnny Brook, Miss E. Rochford and Mrs M. Egerton for allowing their premises and livestock to be photographed, and Alastair Mews, BVM & S, MSc, DipBiol, MRCVS, Alan Slater, BVMS, MRCVS, Gifford Lewis, BSc, MRSH and S. J. Lines, FIMLS, for their helpful and constructive comments on the manuscript.

The authors and publishers also thank the following for providing photographs: Mrs M. Stevens p. 10; Mrs H. Brace p. 11; Mrs R. M. Ragg p. 12; Mrs V. Thornley p. 13; Mrs J. Oliver pp. 54-5, 58; all other photographs were taken by Geoff Goode. The line drawing is by Sue Sharples.

First published in Great Britain in 1982
by Ward Lock Limited, 8 Clifford Street,
London W1X 1RB, an Egmont Company

Reprinted in 1986

Designed by Charlotte Westbrook
House editor Helen Douglas-Cooper
Cover photographs by Marc Henrie

Test filmset in 11 point Baskerville

Set, printed and bound in Great Britain by Netherwood Dalton and Co Ltd.

British Library Cataloguing in Publication Data

Halliday, John
Practical goat-keeping.
1. Goats
I. Title II. Halliday, Jill
636.3'9083   SF383

ISBN 0-7063-6486 4

Contents

Introduction

What could be better in a time of rampant inflation than to pursue a hobby that provides some of the family's food, not only cheaply, but also in a wholesome, unprocessed state? A deep sense of satisfaction can be gained from producing some of life's staples, if only because in doing so, you have gained a measure of independence from the price rises, scarcities and interruptions in supply that are, increasingly, features of modern life.

Goats will provide milk from which you can make other dairy products, such as cheese, yoghurt, cream and butter. Surplus male kids can contribute to the meat supply. Their skins can be cured for use as rugs, or made into articles such as bags and purses. Their manure will be a welcome source of fertility for the garden. Goats are intelligent animals and are responsive and engaging companions. Their size is such that they are relatively easily handled, certainly more so than cows.

Lest this should be too idyllic, let us say straight away that keeping goats is not without its demands and responsibilities. They have to be housed and fed, and well fed if they are to produce milk in economic quantities. They need to be milked twice a day at regular times, irrespective of whether their owners need to recover from a late night or would like to take a weekend away. Special arrangements need to be made if the family wants to take a holiday. Money must be found, sometimes in large amounts, to pay for such things as the year's supply of hay or an unexpected veterinary bill. Although, with good management and care, goats can be expected to be as healthy as any other form of stock, illness and injury can occur, sometimes so severely that death is the sequel or euthanasia the humane alternative.

Some misunderstandings or mis-information about goats still circulate in spite of sterling work by the British Goat Society, so it may be as well to put the record straight at this point. Firstly, all goats do not smell. The male goat does produce a strong odour, particularly during the breeding season, as a means of attracting females. Goats milk does not have a 'strong' taste, if it is produced and kept under hygienic conditions. Goats milk is not 'richer' than cows milk; for all practical purposes, and particularly with regard to butterfat content, the composition of the two milks is similar. However, goats milk is sufficiently different in some of its characteristics for it to be of benefit to people who have difficulties in digesting other milks, and for it to be an acceptable substitute for cows milk in cases of eczema and asthma where these are due to allergy to cows milk and other cow products.

Although goats are inherently productive animals, the keeping of them does not automatically guarantee large profits for their owners. Occasionally, good prices can be obtained for exported animals, but such goats invariably come from quality stock with established reputations in the show-ring and in the form of official milk records. No official agency or marketing organization for the disposal of goats milk and dairy products exists in Great Britain. That is not to say that commercial goat-keeping is not possible in this country, but it does mean that the number of full-time goat farmers is low. Since the would-be goat farmer must find capital to purchase and equip a holding of sufficient size and then find time to farm it, process the milk, package and distribute it, this is not surprising.

Goat-keeping is an absorbing hobby and one that, even on a small scale, can do better than break even financially. It can provide a nutritious part of the family's diet at low cost. Competitive showing of goats is, in itself, a fascinating activity and goat-keepers are amongst the friendliest and most helpful people you could hope to meet. This book, we hope, will serve as a practical introduction to the world of goats and goat-keeping.

Introduction (second edition)

In the time since this book was first published, there have been significant developments in goat-keeping. In the first edition we noted the foundation of the Goat Veterinary Society. We are pleased to see that this has become firmly established to the benefit of veterinarians goat-keepers and goats themselves. The formation of the Goat Producers Association, the increasing number of commercial goat herds and the serious treatment of goats in the farming press show that the goat is gaining its rightful place in British agriculture, at last. On the debit side, the spread of caprine arthritis encephalitis and the reduction in already inadequate resources applied to the science of goat-keeping by government agencies are matters of regret.

This book deliberately concentrates upon the principles and practice of goat-keeping; we trust that it will continue to introduce newcomers to this fascinating pursuit and to form a basis for a lifetime of enjoyment.

1 The goat: origin and background

History

Archaeologists date the earliest goat remains in the Pliocene era (from one to eleven million years ago). In Neolithic times (3,000 to 6,000 BC), goats were amongst the first animals to be domesticated, when man began to turn from hunting to an agrarian existence. Evidence of sheep and goats has been found at Neolithic sites in many parts of the world.

Classification

Goats belong to the order *Artiodactyla* (hoofed animals with two toes), within which they are classified in the family *Bovidae*, which includes cattle, sheep and deer. The genus *Capra*, which is widely distributed throughout the world, includes the true goats, the ibexes and the markhors. The domesticated goat is known as *Capra hircus hircus*, whose ancestors are *C.aegagrus* of Greece and Asia Minor, *C.falconeri* of the Himalayas, and *C.prisca* of the Mediterranean region.

Up to the middle of the last century, the indigenous goats of Britain were distinguishable as three separate types, the English, the Welsh and the Irish.

The English goat had a shaggy blue/grey coat. Both sexes were bearded and had horns that swept up, back and outwards in a smooth curve. Little information on milk yields is known. Attempts have been made in recent years to revive the breed, but it seems unlikely that any of the original English stock exists, although animals of similar appearance occur.

Welsh goats have long white curly coats. Their horns point straight backwards and do not sweep outwards as in English goats. The best known herd in captivity is the Queen's herd at Whipsnade Zoo, from which regimental mascots are obtained.

Irish goats are usually black and white, with horns rising straight upwards and slightly backwards in the male, but short and pointed in the female.

Feral and wild goats

Feral goats exist in Britain today, descended from the original native stock, though some of them show evidence of Swiss blood since they descend from 'escapes'. Goats, much more so than cattle or sheep, revert to the wild and cast off domestication easily. Wild goats are found in the isolated mountainous parts of the British Isles, principally in the Highlands and Islands of Scotland, in Northumbria, in the Welsh mountains and in Ireland. Wild herds are at risk as moorland is reclaimed for forestry. The goats cause serious damage to young trees and consequently are considered pests by foresters, and shot.

Registration

Apart from one or two minor attempts, no registers of English, Irish or Welsh goats have been kept. The British Goat Society was formed in 1879, to cater for the upsurge in interest in goat breeding. The Herd Book was opened in 1886, but there were insufficient numbers of any type to justify opening Breed Sections until some years later. The progeny of goats that were accepted for Herd Book registration were known as 'Anglo-Nubian-Swiss'. This term was later changed to 'British', and is still used today to describe goats of mixed origin that cannot be classed in breeds.

A breed section for Toggenburgs was opened in 1905, for Anglo-Nubians in 1910, for Saanens in 1922 (to coincide with the first major import), and for the British Alpine, British Toggenburg and British Saanen in 1925. The first Golden Guernseys were registered with the British Goat Society in 1971, and the English Guernseys in 1977. However, the island of Guernsey opened its own Herd Book in 1922 for registration of Golden Guernseys.

Goats in world agriculture

On a worldwide scale, goats are used primarily for the production of meat, with milk, skins and hair of secondary importance. One in seven of all grazing animals in the world are goats. In Britain and other temperate countries, where their main use is for milk production, the population is comparatively small. However, in many tropical countries where meat is the main product, they are often the most important species of livestock kept. In these countries skins become an important secondary product.

Only the Cashmere and Angora goats produce hair suitable for commercial use, and whilst the industry is small, it is economically important in the regions where climatic conditions are favourable.

Goat-keeping in Britain

In Britain, goats are kept almost exclusively for milk production, and although surplus kids are used for meat, this tends to be for domestic consumption and no organized industry exists. With the recent upsurge of interest in smallholding, goats have increased in popularity. Whilst by no means all of these goats are registered with the British Goat Society, the number registered continues to rise annually. No population figures are collected for the total number of goats in Britain, but the number registered annually exceeds 7,500.

The vast increase in registrations has stimulated the formation of breed societies. Although these do not handle registrations, they foster interest in the breeds and encourage true breed types. Golden Guernsey and Anglo Nubian breeders were the first to form such groups, but most of the other breeds now have their own societies.

A few large commercial herds exist in Britain which sell milk and milk products. Most goats are kept as back yard and smallholding livestock. Excess produce may be sold to provide a small income, but principally the animals give milk for the family and are treated as productive 'pets'. For this reason, little scientific research is carried out on either the nutrition or diseases of goats. Such work as has been done in this country has tended to use goats as a convenient experimental species rather than for the sake of the goats themselves. Hopefully, with the recent formation of the Goat Veterinary Society, this situation may begin to change.

2 Breeds of goats

Anglo-Nubian

The Anglo-Nubian, perhaps the most distinctive of British breeds, is derived from indigenous goats crossed with Zariby goats from Egypt and Jumna Pari goats from India, imported at the turn of the century. Sufficient goats of a suitable type had been bred by 1910 to merit opening the Anglo-Nubian section of the Herd Book. Although the numbers of Anglo-Nubians were low for many years, great improvements have been made in the productivity and conformation of the breed. In the 1970s the numbers registered were exceeded only by British Saanens.

The Anglo-Nubian has a characteristically 'eastern' appearance, with a convex Roman nose, and lop ears. Undershot jaws occur frequently, but can only be considered a fault if the teeth are visible when the mouth is closed. The head is carried high and the neck is free from tassels. The legs are long in proportion to the depth of body, compared to other breeds, and are very straight with no tendency to 'cow hocks'. The pasterns are normally strong to give an upright gait. The withers and point of the hip bones are high, giving a slight curve to the backbone, which should not be so pronounced as to become a dip. Twisted facial bones giving a wry nose and twisted jaws are faults. Pendulous udders and large teats occur all too frequently. Coat colour is very variable, from black through all shades of brown to pale cream and white, plain or dappled.

Milk yields of Anglo-Nubians tend to be slightly lower than those of most of the Swiss breeds, although individuals have given 365-day lactations approaching 2,000 kg. An average animal should yield 1,000 kg and butterfats usually range from 4% to 6%. The range of protein levels recorded is 2·40 to 4·12% and appears to correlate directly with butterfat levels.

Saanen

Saanen goats were imported from the Saane and Simmental valleys of Switzerland. A few animals came into Britain around the turn of the

Anglo-Nubian: R121 Salvington Hecuba Q*1 Br. Ch., AN7676H.

century, but the first major importation was in 1922 from Holland, with a further import in 1965 from Switzerland. On a world scale, they are probably the most popular, both as a pure breed and for cross breeding to produce breeds such as the British Saanen and the American Saanen. The Breed Section in the British Herd Book was opened in 1922 for the twenty-nine goats imported from Holland. The breed has been kept closed, and although this causes problems when trying to select unrelated stock for breeding, high yields and good conformation have been maintained. There is a tendency to weak hocks and pasterns, and much effort is made to breed these faults out. In Switzerland the breed is described as hornless, but, since breeding polled animals together results in a high incidence of hermaphroditism, it is unclear how this is achieved. It is reported, however, that a high proportion of horned kids are killed at birth. Between 100 and 200 Saanens are registered each year.

In appearance the Saanen is pure white, with shorter legs in proportion to body depth than the British Saanen. The face is slightly dished with ears erect and pointing forward. The coat is short, but fringes of long hair often occur. The back should be straight with no tendency to a hump, referred to as a 'roach' back.

Milk yields are generally good with 365-day lactations of 1,000 to 1,500 kg frequently recorded. Butterfats are in the region of 3% to 4%.

British Saanen

This breed was formed early this century by crossing imported Saanens with indigenous goats. By 1921, sufficient goats of the type had been bred to merit its recognition as a separate breed. In 1925 the Breed Section of the Herd Book was opened. This has always been an open book thereby allowing stock to be upgraded. Numerically, the breed is the most popular, with around 1,000 animals being registered each year.

The British Saanen is bigger than the Saanen, and because it is an open breed, it has been possible to make more attempts to breed out weakness of the legs. The coat is pure white and short. The facial line is straighter than that of the Saanen.

Some of the highest milk yields are given by goats of this breed. Lactations of 1,000 to 2,000 kg are frequently recorded and yields of over 2,000 kg in 365 days have been given by exceptional animals. Butterfats are generally in the 3% to 4% range.

Toggenburg

The Toggenburg is a Swiss breed originating from the regions of Ober-toggenburg and Werdenburg. Importations into Britain have occurred

A Breed Champion Saanen Milker: R122 Hetherton Aphrodite *1 Br. Ch.

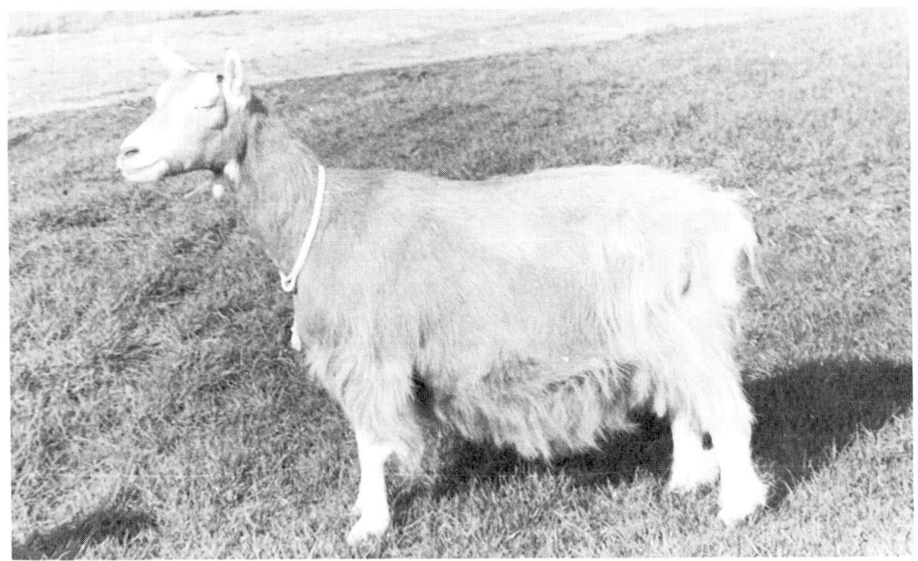

Toggenburg: Tanatside Trina *T.1742P.

from as early as 1884. Animals were brought in with the 1922 and 1965 importations of Saanens but unfortunately those that came in with the later one did little to improve milk yields.

As with the Saanens, attempts have been made to import further stock, but stringent import regulations have prevented them. The Toggenburg breed section of the Herd Book was opened in 1905, and the breed enjoyed great popularity early on. However, other breeds overtook the Toggenburg due to their better milk yields and butterfats. Toggenburgs are extremely good milkers in other countries. The problem appears to lie with the stock in Britain rather than the breed itself. About 100 animals are registered each year, but whilst the numbers are small, the breed serves as a genetic pool of true breed type for use in the British Toggenburg breed.

In appearance, the Toggenburg is a small, compact animal, sturdy with strong pasterns and a good capacity for bulk food. The colour is fawn to brown with white 'Swiss' markings. The coat nearly always has some long hair, which may be restricted to a fringe down the back and hind legs or cover the whole body except the head and neck. The Toggenburg is a hardy little animal and thrives well on extensive browsing and grazing, producing milk quite economically.

Occasional exceptional milkers occur in the breed, but few achieve much more than 1,000 kg in a 365-day lactation with butterfats of 3% to 4%.

British Toggenburg

The British Toggenburg arose following the use of pure Toggenburg males on animals of mixed descent in the early years of the century. By the 1920s the type was established and it was recognized by the opening of the Breed Section in 1925. The breed has always enjoyed great popularity with 700 to 800 animals currently being registered annually.

The British Toggenburg is a larger, rangier animal than its purebred relative and in colour varies from the fawn of Toggenburgs to a chocolate brown, with white 'Swiss' markings. The coat is usually fine and short, without fringes of long hair. The face is usually straight rather than dished as in the Toggenburg.

The yields of British Toggenburgs are frequently high, with many goats giving 1,000 to 1,500 kg in a 365-day lactation. Yields between 1,500 kg and 2,000 kg are not unusual. Butterfats tend to range between 3·5% and 4·5%.

British Alpine

The British Alpine is a breed developed in Britain. No equivalent Swiss breed exists and it is certainly very different from most French Alpines.

British Alpine: CH. RM. 34 Sundial Sarah Q*2 Br. Ch., BA.7650H.

The striking black-and-white appearance was established in the early years of this century, and derives, it is thought, from a goat called Sedgemere Faith and her progeny. Sedgemere Faith was imported in 1903 and was black with white 'Swiss' markings. The breed was defined in 1919 and in 1926 a Breed Section was opened.

For many years the breed enjoyed enormous popularity with several well-known herds producing extremely good high-yielding animals. During the 1970s, the breed suffered a decline in quality and performance and a consequent decline in popularity. With no pure breed to cross out to, to re-establish breed characters, British Alpine breeders are experiencing considerable difficulties in regaining good udder shape and maintaining yields without losing the pure black coat colour. About 250 animals are currently registered each year.

A good British Alpine is a striking animal with a pure black coat and white 'Swiss' markings. The coat is short and the animal is frequently long legged and rangy. It enjoys extensive browsing and grazing and has a big capacity for bulk foods. Occasionally yields approaching 2,000 kg in a 365-day lactation occur, but more frequently the yields are 1,000 to 1,500 kg, with butterfats of 3% to 4%.

Golden Guernsey: Ashrose Gaynor second kidder, sire Novington Boris, dam Alderkar Gaynor.

Golden Guernsey

The history of the Golden Guernsey goat is unknown until Miss Milbourne of L'Ancresse began her life's work of preserving the breed. The Island Herd Book was opened in 1922 so a long history of pedigree was established before importations into Britain began. A Golden Guernsey Register was opened by the British Goat Society in 1971. The breed is steadily increasing in popularity, and, by judicious selection, improving in conformation and yield. Around forty goats are registered each year.

The Golden Guernsey is a uniform golden colour, which ranges from almost cream to ginger. The coat may be short and fine, but nearly always has long fringes or sometimes long hair over the whole body from the shoulders backwards. The goat is short and small, with a tendency to weak hocks.

Lactations occasionally reach 1,000 kg but are frequently lower, with butterfats in the region of 4% to 5%. However, these yields are achieved ideally on low concentrate, high roughage diets and the Golden Guernsey may well come to find a niche as a hardy animal suitable for smallholders.

English Guernsey

The English Guernsey is the progeny of Golden Guernsey females out-crossed to Saanen or British Saanen sires, and the stock from these crosses mated to Golden Guernsey or English Guernsey sires for three generations. The English Guernsey Register was opened in 1975 by the British Goat Society and the first registrations are now being made. Because of the British Saanen blood in the breed, the animals are often a paler cream colour rather than the darker golden colour of the Golden Guernsey. Hopefully, the use of British Saanen blood will improve the Golden Guernsey stock, both in conformation and productivity.

British

British goats are those goats not eligible for registration in a breed section. They may be cross-bred goats or goats of mixed ancestry. Many high-yielding, good-quality animals are British. The ability to register goats in the Herd Book or Registers as British is beneficial in two ways. It allows, firstly, improvement of the breeds by crossing out to other breeds to fix characters, and secondly, the upgrading of unregistered stock.

The goats may favour any of the breeds in appearance or be a mixture of features.

3 Selection of stock

Considerations before purchase

There are various questions that the intending goat-keeper must ask himself before buying goats. First, are the premises adequate in size and suitably located for livestock? Goats can be kept in a relatively small garden, provided that suitable housing and exercise areas are available and that muck heaps do not cause a nuisance to neighbours. A certain amount of space for hay and straw storage is also necessary.

Neighbours must also be considered. Goats are not noisy by nature, but at feeding times, and when they are in season in the winter, they will be noisy. An irritable neighbour may take exception to the noise, and if the local council takes his side, the goats may have to go.

Goats require attention at least twice a day, seven days a week, and consequently can considerably limit a person's activities. Apart from milking and feeding, time is required each day for handling the milk, and making any dairy produce, and each week time must be set aside for routine jobs such as mucking out, hoof trimming and running repairs to buildings and fencing.

Finances are important. Feed must be paid for and hay and straw has to be bought whether milk is being sold or not. Equipment on the whole is not costly, but nevertheless, certain items will require repair or replacement. Vets bills, hopefully occasional, seem never to be small and often to be unavoidable. Adequate housing is necessary. Frequently existing buildings can be adapted at a fairly low cost, but in some cases, a substantial outlay may be essential.

Numbers, sexes and ages to buy

Goats are herd animals, and it is unnatural for them to be kept on their own. Occasionally, animals will settle without companions, but on the whole a goat on its own will be noisy and troublesome, as it will spend its time trying to reach human company. A minimum of two female goats should be kept. Each goat can be mated in alternate years to give a continuous milk supply.

There is no need for the household goat-keeper to keep a male goat although some information on males is given in Chapter 6. Castrate males can make adequate pets but provide no useful return.

Each age group has advantages and disadvantages from the buyer's point of view. Kids quickly become accustomed to their new owner's ways, but are expensive to rear and do not come into production for a long time. Goatlings also settle easily; the costly part of rearing is over, and they come into production sooner. However, a novice goat-keeper may well have difficulty getting a first kidder used to being milked.

A goat in milk will take longer to settle into its new surroundings than a younger animal, but is immediately productive and, hopefully, easier to milk. When buying a goat in milk it is advisable to try milking it before confirming the purchase.

Conformation

There are several points to look for when buying a milch animal. It is essential to consider those connected with milk production rather than decorative or 'fancy' points.

A milk-type goat should be wedge shaped when looked at from the side and from above.The widest part of the wedge represents the

Conformation—what to look for.

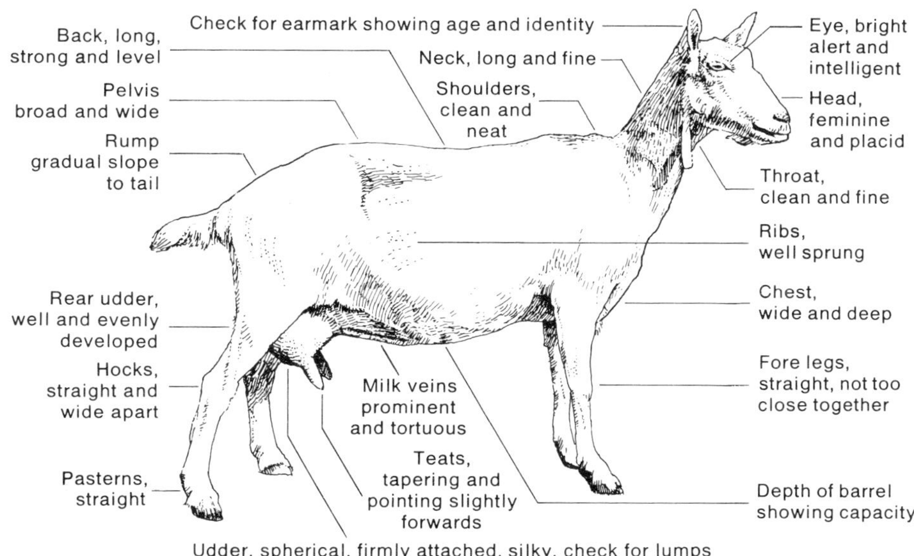

rumen capacity and indicates that the goat has a good capacity for bulk foods. The back should be strong and firm, with no tendency to dip behind the shoulders. A weakness here may well shorten the productive life of the animal since the back has to support the weight of large amounts of food and developing kids.

As with any grazing or browsing animal, the feet must be sound to enable the goat to seek out her food easily. 'Cow hocks', i.e., the hocks of the hind legs turning in towards each other, and weak pasterns again may shorten the productive life of the animal, the former because it can result in damage to the udder, the latter because the pasterns carry a substantial weight load.

To last for an economic number of lactations, the udder should be as near spherical in shape as possible, not pendulous, be well attached to the body, and should be soft and supple to the touch. Lumps in the udder tissue may be indicative of previous mastitis infections and should be viewed with concern, or they may be cystic tissue. Teats should not be so small that they cannot be milked easily, nor should they be so large that they make milking difficult, suckling impossible and are liable to injury.

Health

An experienced stockman will always be able to tell whether an animal is fit by its alert expression, bright eyes, healthy coat and general bodily condition. An animal in poor health and condition should not be purchased.

Where to buy

The most reliable sources of information regarding quality stock for sale are advertisements in British Goat Society publications, and the local goat-keeper's organizations. Secretaries of some Affiliated Societies keep lists of goats for sale, and can give invaluable help to the beginner. Local newspapers sometimes advertise goats in their classified columns. Although occasionally animals of reasonable quality may be obtained in this way, such advertisements are best ignored by the beginner. A novice might well consider enlisting the help of an official of the local society when making his first purchase.

Animals registered with the British Goat Society are usually a better investment than unregistered goats. The seller should be able to produce the registration card, and although this is not given to the purchaser with the goat, it eventually comes into his hands when it has been through the British Goat Society's office.

18

4 Buildings, equipment and control

Why housing is needed

Goats have relatively thin coats with neither the insulating properties of a thick fleece nor, since most of the fat they carry is within the body cavity, those of a layer of fat beneath the skin. In addition, the coat does not have the same water-repellent property as sheep's fleece, due to the absence of lanolin. As a result, they are intolerant of wind and rain. The drop in yield of grazing goats is marked in periods of damp and cold windy weather. It is usual in cool climates to house milkers at night all year round and all goats through the winter months. In favourable summer weather there is no reason why older kids and goatlings should not remain out at night where they have access to adequate natural cover or to field shelters.

Apart from protecting stock from climatic extremes, housing has other advantages. Firstly, it permits the animals to be more easily controlled, for example when concentrates are to be fed or routine tasks such as milking, hoof trimming and grooming are undertaken. Secondly, it lets the owner perform such tasks under cover. Thirdly, it allows easier collection of manure for subsequent use in the vegetable garden or for forage crops.

Requirements of housing

The main requirement is that the housing should be wind and waterproof. It must also provide adequate ventilation to remove stale air. (Condensation on windows or other structures indicates that ventilation is not sufficient). Poor ventilation can result in an increased susceptibility to pneumonia, but ventilation must not expose the animals to draughts.

There is no need to provide heating. Unlike simple stomached animals, such as pigs, ruminants carry fermenting roughage in their rumens as an additional heat source. Goats will remain comfortable in the coldest of conditions if they are given deep, dry bedding and are protected from draughts.

Features of housing

The following features should be taken into account when planning to build new housing or adapt existing buildings for goats.

Site

The ideal position for a goat house is facing south or south-east on a gentle slope so that cold air drains away from it. Avoid hollows, which can be frost pockets. Houses under trees can obtain some protection from sun in midsummer but are often cold and damp for most of the year, risk being damaged by falling branches, and are best avoided.

Storage

Bedding and foodstuffs are bulky. Storage barns or sheds should be close to the stock, to reduce time and effort.

Access

A hard road to goat-house and barn will be of advantage for deliveries of foodstuffs and transport of animals. Access to exercise yards or grazing and to the muck heap needs to be considered.

Services

Electricity for lighting is highly desirable since in winter, at least, most of the routine daily tasks will probably take place in the dark. In the absence of a mains supply, a lighting system powered from a 12 volt car battery would be preferable to anything else. Other forms of lighting using paraffin or liquefied petroleum gas, should only be considered as a last resort and then used with extreme care.

Position lamps so that they are away from combustible materials, impossible to knock over, out of draughts and out of the reach of children and animals.

An adult doe in milk will drink around 18 litre (4 gall) of water each day. Since this volume weighs about 18 kg (40 lb), a water supply close at hand will be of obvious benefit. Limited amounts of water will also be required for washing down pens and passageways after mucking out, and for cleaning in the milking area and dairy. An inside tap will be less likely to freeze up in mid-winter. A hot water supply, whilst being necessary for the dairy, is not essential in the goat-house, although goats do prefer warm drinking water.

Drainage is not required from goat housing. Goats do not produce great volumes of urine, and most of this will be absorbed by the bedding to produce first class manure for the garden. However, proper provision must be made for foul drainage from the dairy. The Building Control Officer from the local District Council will be pleased to advise. Dairy effluent, which contains milk, detergents, sterilizing chemicals and so on, should not be released untreated into water courses or drainage ditches.

Nearness to dwelling

Vandalism and theft of livestock are, regrettably, increasing problems in many areas. Goats require attention routinely whatever the weather and, for example, at kidding time may need to be visited in the middle of the night. For convenience and peace of mind, therefore, the nearer the goat-house is to the dwelling, the better.

Before commencing building

Most goat-keepers interested in providing a supply of milk and dairy products for the house will find it difficult to justify new housing because of the costs of materials and labour. If, however, the goat-keeper has the necessary skills and resources and is prepared to put up new buildings, the Planning and Building Regulations should be checked with the local Council before starting to build. These officials will be a great source of help and advice. Planning permission is not likely to be required where it is intended to erect a 'temporary' building (e.g., one of sectional, timber-frame construction, that is not on permanent foundations), and where the enterprise is not to be conducted on a commercial basis. A tenant would be wise to obtain his landlord's permission before embarking on a goat-keeping venture. Similarly, an owner-occupier should check his deeds for restrictive covenants.

Construction

Most goat-keepers will probably adapt existing buildings, rather than put up purpose-built ones. The following points are written with this in mind, although the general principles are also applicable to the design of new buildings.

Roofs

The principal requirement is that the roof should be weatherproof and, subject to this, more or less any material is suitable. However, the thinner materials (such as asbestos or corrugated steel) are not satisfactory unless either they are lined with an insulating material, or sufficient draught-free ventilation is provided to ensure that the temperature within the house is the same as that of the external air in winter, and that increased air flow can be provided to minimize the effects of excessive solar gain in summer.

When warm air rises from the goats and strikes a cold surface, such as uninsulated sheeting, the cooling results in a convection downdraught, which is usually directed back on to the animals. Ventilation (e.g., from hopper windows at eaves level) will remove the warm air before it rises to the underside of the roof.

Because of the difficulty of providing adequate ventilation without creating draughts, it would be safer to insulate a thin roof, and settle for less ventilation. In such conditions, a vapour barrier of impervious material, (e.g., polythene sheeting) should be provided between the insulation and the air within the house to avoid condensation and subsequent impairment of the insulating material's efficiency.

Floors

Floors for pens and passageways must be easy to clean. Although earth floors can be successful on free-draining sands or gravels, on clay soils they become waterlogged and are attractive to vermin, particularly rats. A solid surface is preferable. Wooden floors are not suitable. Brick floors, found in old stables and outhouses, are ideal and should not be removed if such buildings are converted. In new work or extensive renovations, concrete will be the most suitable material. Its major disadvantages are that it is cold and expensive. Little can be done about the latter, but its heat-retaining characteristics can be improved in several ways. Firstly, a damp-proof membrane should be used beneath the floor to prevent the soil moisture soaking into the concrete. (Polythene sheet is perhaps the most suitable material to use as a membrane. It should be laid on top of a layer of well-rammed hardcore, which in turn has been blinded with sand or sieved ashes to avoid any projections that might puncture the membrane). Secondly, an insulating material can be incorporated beneath the floor slab. Any material that is rigid, impervious and traps still air, will serve. Polystyrene and proprietary insulating boards are suitable, as are such things as old bottles and jars, drain tiles or hollow blocks. When using some of these materials, it may be more satisfactory to cast a sub-floor in concrete, lay a membrane on top, then the insulation and finish with a sand/cement screed about 50 mm (2 in) thick. Thirdly, although not a feature of construction, whatever the nature of the floor, the animals should not be expected to lie directly on it. A layer of bedding, usually straw, at least 150 mm (6 in) thick, should be provided. A layer of sawdust, peat or shavings under the straw will further insulate the animals from the ground and help to absorb liquid manure.

Concrete floors 75 mm (3 in) thick will be adequate for pens and passageways that are not likely to carry heavy machinery. A mix of 1:2½:4 (cement:sand:coarse aggregate) should be used for concrete and 1:3 (cement:sand) for screeds, which should be laid very dry. Floors should be well tamped and finished with a wood float to give a smooth, non-slip surface. Drainage channels are unnecessary, although the whole floor can, with advantage, be laid with a fall of 25 or 50 mm (1 or 2 in) from the back of each pen to the front, to aid the disposal of water used to wash down pens after mucking out.

Walls

The same general considerations apply to walls as were stated above for roofs, that is, they must be weatherproof and provide a measure of insulation. For these reasons, corrugated iron is not suitable and corrugated asbestos is best avoided where it is liable to be damaged by livestock. Both these materials would be improved if lined with an insulating material that is resistant to vermin, urine and physical damage, and not toxic to goats. An internal lining of, for example, exterior-grade ply would be most suitable and should protect any insulation.

For new work, 150 mm (6 in) or 225 mm (9 in) hollow concrete blocks (or load-bearing insulating blocks) should provide the ideal combination of structural stability with a measure of insulation. These materials will require rendering or painting (with a masonry paint) to avoid penetrative damp. Internally, render or, if smooth blocks and flush joints have been used, a painted finish will assist in maintaining cleanliness. Brickwork is expensive and therefore unlikely to be used when building from new. If faced with the situation where it is unavoidable (e.g., a Planning Department requirement to match existing buildings), cavity wall construction will be the most satisfactory. Solid 225 mm (9 in) brick walls will give less insulation, but are probably not worth attempting to improve. Walls of one brick thickness will need insulating and lining. All brick or block walls must have an adequate damp proof course to prevent rising damp.

Interior layout of goat housing

The intending goat-keeper will, more than likely, have to work within the confines of an existing building and adapt this to his or her needs. There seems little point, therefore, in providing sets of plans for 'ideal' goat housing, and we will continue to concentrate upon broad principles.

Essentially, there are two ways of laying out the goat-house. First, a separate pen can be provided for each animal. Second, goats can be loose housed, that is, housed together.

Separate pens

Separate penning is by far the commonest system used in Britain, par-

The interior of a well laid-out goathouse, showing individual penning.

ticularly where smaller numbers of goats are kept for domestic milk production. Under this system, each milking animal is given a pen of about 2·25 sq m to 2·75 sq m in area (25 sq ft to 30 sq ft). Ideally, pens should be oblong rather than square (say, 1·8 m by 1·2 m [6 ft by 4 ft]) with the door in one of the short sides. Such an arrangement allows the goat more room for lying down, kidding, and so on.

The principal advantage of this system is that it allows animals to be fed individually for optimum milk yield, and prevents bullying. Its disadvantage is that it requires a larger outlay on materials for pen divisions than loose housing.

It is usual practice to provide communal pens for different ages of young stock, i.e., a pen for goatlings and a pen for kids. About 1·4 sq m (15 sq ft) per animal should be allowed for older kids and goatlings.

Loose housing

The principal advantage of loose housing is the saving in cost of construction due to the few internal divisions that are used. However, where few animals are to be housed and where the construction work is to be carried out by the goat-keeper, perhaps using reclaimed materials, such cost savings are minimal. Loose housing is probably of interest to the herd owner contemplating a commercial or semi-commercial project. With a loose-housed herd, provision must be made for goats to receive individual concentrate rations (e.g., by means of a heck — a fence with holes through which goats put their heads to reach buckets on the far side, and which has some means of locking so that they cannot remove their heads until released by the herdsman when all have finished eating).

Individual pens must be provided for kidding and sick goats. Groups of goats of different ages must be housed separately.

An allowance of 1·4 sq m (15 sq ft) per goat would be adequate for loose-housed milkers. For goats of the larger breeds, up to 2.0 sq m (21 sq ft) would be preferable.

Pen walls

There is a wide choice of material for constructing internal walls. Concrete blocks 100 mm (4 in) thick, or single brickwork, are ideal. Wood can be used, but has the disadvantage that it will rot where it is in contact with damp bedding and must therefore be kept at least 150 mm (6 in) off the finished floor level. Plywood 12·5 mm (½ in) thick can be used, as can tongued and grooved boarding fixed vertically to wooden framing at least 50 mm by 50 mm (2 in by 2 in) in section. Wire netting or chain link fencing is less suitable for partitions. Both tend to sag after a time as goats will always stand on them, and with these and weldmesh there is the risk that an animal will trap one of its legs. Solid materials will help to cut down draughts. Wooden partitions can be made demountable and thereby allow some flexibility to take account of the varying numbers of kids, goatlings and milkers in the herd. All pen divisions should be at least 1·2 m (4 ft) high.

Doors

All doors in the goat-shed and to the outside should open outwards, and doorways should be at least 750 mm (2 ft 6 in) and preferably 900 mm (3 ft) wide. Doors should be hung so that the tops are the same height as pen walls.

Doors can be made of metal (e.g., galvanized sheet on a framework of steel tube), but for the home handyman, timber is probably more suitable and can be in sheet form (e.g., plywood) or framed boarding as suggested earlier for pen divisions. A pair of strap (or T) hinges at least 300 mm (1 ft) in size is necessary to provide the strength required to support the door and the weight of a goat standing on the framework. Barrel bolts are usually adequate to secure doors, but occasionally a goat will learn to open these, in which case pad bolts can be substituted. To foil the confirmed escaper, pad bolts can be further secured by means of a clip through the hole where the padlock would normally go.

Passageways

Access passageways should be at least as wide as the doors that open into them. If space permits, the wider the passageway that can be provided, the better. A passage 1·4 m to 1·8 m (4 ft 6 in to 6 ft) in width will allow adequate space for manoeuvring barrows when cleaning out pens. Always arrange for each pen to open directly off the passage. Interconnecting pens are an unmitigated nuisance.

Fixtures and fittings

Hayracks

Since goats are fastidious feeders, means of holding bulk foodstuffs must be provided. The most satisfactory arrangement is to fit a rack in each pen. Haynets are a poor alternative. They are time-consuming to fill, they have caused the death (by hanging) of innumerable kids, and are best avoided.

Racks can be made from timber or metal. The essential requirement is that the front slats should be 50 mm (2 in) apart. Set closer, goats will find it difficult to feed, and wider spacing will allow feed to be wasted.

Overall dimensions of each rack can be varied to suit individual situations, but we have found the following useful for single pens: height 600 mm (2 ft), width 710 mm (2 ft 4 in), depth 300 mm (1 ft) at the top tapering to 150 mm (6 in). About 450 mm (1 ft 6 in) linear run of hayrack per goat should be sufficient in communal pens. Racks should be hung so that they are at least 750 mm (2 ft 6 in) clear of the floor to avoid the contents becoming fouled.

A lid will reduce wastage of hay, kale, and so on by preventing the animals dragging it out from the top of the rack. Wastage can be further reduced, where pens are sufficiently large, by means of the Egerton rack. This consists of a second set of slats 165 mm (6½ in) apart, 450 mm (1 ft 6 in) in front of the rack and parallel to it. These slats run from a point level with the top of the rack to the floor and allow the goat's head and neck through, but not its shoulders. Any hay pulled from the rack that falls to the floor will not be trampled upon, and will be eaten. A further refinement of this design is to have a means of access to remove any hay from beneath the rack, so that it can be collected and fed to young stock.

Racks should be hung on the front or side walls of each pen so that they can be filled from the passage.

Bucket holders

Water, concentrate mixtures and other feedstuffs, such as roots, sugar beet pulp and vegetable and fruit trimmings, are fed most conveniently from buckets. To avoid these being fouled or damaged by the goats, and to save time, these should be outside the pen. A hole 300 mm (1 ft) high and 250 mm (10 in) wide in the door or front wall of the pen will allow the goat to put its head and neck through to reach buckets hanging on the passage side. (The bottom of the opening should be about 600 mm [2 ft] above the floor). Drop-ring bucket holders can be fashioned easily from 6 mm (¼ in) mild steel rod bent to the circumference of the bucket and secured to two large screw eyes. Such holders fold down when not in use and thus do not obstruct the passage. As an alternative, heavy-duty vee-belting fixed by staples can be used, which will not impede movement of people or barrows.

The Egerton hayrack, which consists of two parallel rows of slats, reduces the amount of hay that gets wasted.

Exercise yards

Where grazing is limited, far better use can be made of the space by growing crops to provide a succession of bulk foods throughout the year. In these circumstances, a yard is essential for exercise. Ideally, such a yard should be of concrete laid to a fall to shed rainwater, and finished to provide a non-slip surface. The yard should be fenced to a height of about 1·5 m (5 ft). About 1·4 sq m (15 sq ft) should be allowed for each milker. Some simple form of protection from wind and rain should be provided.

Wall-mounted or free-standing racks for bulk foods, such as hay, kale, lucerne and grass, should be provided. A free-standing rack can be constructed using a pair of X-shaped end-frames with slats, bars or welded mesh panels (with openings at least 75 mm by 50 mm [3 in by 2 in]) on the feeding faces. About 450 mm (1 ft 6 in) linear run of feeding face should be allowed per goat. A manger mounted beneath the rack would be useful for feeding roots, apples and so on.

Ball-valve controlled water troughs will save considerable time spent carting water buckets. Water pipes should be laid at a depth of about 1 m (3 ft) and insulated where they rise above this to prevent freezing-up in winter. An added refinement would be to provide thermostatically-controlled heaters in the troughs. Water troughs must be fitted with drain valves to permit regular twice-daily cleaning.

A covered exercise yard.

Where soil is heavy and inclined to poach, yards provide ideal accommodation during wet weather and through the winter.

Other equipment

Buckets are the usual choice for feeding concentrates and for holding drinking water. Plastic, 9 litre (2 gal) buckets are ideal and will last far longer than galvanized ones, if kept outside the pens as described above. Buckets for food and water should be used for no other purpose.

Amongst the equipment required for cleaning out pens, will be a wheelbarrow (a rubber-tyred builder's barrow is best), stiff brooms, shovel and forks. Round-tined manure forks are far better than square-tined digging forks.

Transport

Usually goats will travel well in vans, or in the back of the family car with the rear seat removed and sacks of straw used to soften hard edges. However, a light trailer is an excellent investment, particularly if you intend to move goats around frequently (e.g., to a stud male for mating, to shows, etc.) or to carry bulky materials (hay, straw, etc.). Worthwhile savings can usually be made if you are prepared to collect such materials from farmers.

Storage

Provision must be made for storage of adequate amounts of hay, straw and other feedstuffs. Whilst concentrated feeds are usually available in small amounts from merchants and hence require relatively little space, hay, straw and roots are bulky and often only freely available in the late summer or autumn. It is therefore useful to have sufficient space to store a full year's supply. It is difficult to give precise indications as to the amount of hay required since this will vary depending on the pattern of management and on other feeding resources, such as grazing and crops.

As a rough guide, a milking goat will consume between ½ and 1 ton each year, which will require 10·5 cu m (125 cu ft) to 21 cu m (245 cu ft) of storage capacity. Similarly, quantities of straw for bedding will vary dependent upon, for example, the intensiveness of the system of management and the frequency with which pens are cleaned out. A generous annual allowance for straw would be around ½ ton per adult, which will occupy about 13 cu m (153 cu ft).

Buildings for storage can be cheap and simple, providing that they keep materials dry and permit some movement of air around the bales. Concrete floors are an advantage though not essential. Bales should always be stacked on bricks, blocks or lengths of timber to hold them off the ground. Polythene or tarpaulin covers should be avoided except as temporary measures. They are inconvenient to use, do not allow stored materials to breathe (and hence result in dampness and subsequent mould growth), and are inclined to rip or blow off in gales.

Concentrates should be stored under cover in bins or containers that are free from damp, vermin proof and have lids that cannot be easily removed by inquisitive goats. In view of the disastrous consequences that can follow from a goat gorging itself at the food bins, they are best kept in a separate building to that in which stock is housed.

Buildings for milking and handling milk

Unlike the dairy cattle farmer, the goat-keeper is remarkably free from rules and regulations concerning facilities for handling and processing milk. Such requirements as there are only apply to milk produced for sale. If the intention is to supply only the family, then no special arrangements are needed beyond a constant attention to hygiene.

Whether two animals or one hundred are being milked, the principles involved are essentially the same. First, the milk must be drawn from the animals, second it must be strained, cooled and stored, and finally, in some cases, it will undergo further processing (e.g., packaging for sale, processing into cheese and yoghurt). The majority of goat-keepers undertake milking in the goat-house (or an adjoining building) and handle and process milk in their kitchens. A more acceptable arrangement, where the scale of the enterprise can justify it, would be to have a milking parlour, separate from the animal's housing, with a dairy adjoining.

A milking parlour need not be elaborate, providing that it is used for no other purpose. The ceiling, walls and floor should all have washable surfaces and the floor must be laid to drain. All drains should be trapped, preferably outside the building. There should be a minimum of ledges, shelves or other horizontal surfaces to collect dust. Coved or rounded junctions where walls meet or where they adjoin the ceiling or floor will aid cleaning. A concrete plinth, topped with non-slip tiles can be cleaned more easily and harbour fewer organisms than the conventional milking stand constructed of rough-sawn second-hand timber.

If a separate parlour is not to be provided, the minimum acceptable arrangement is to have a separate milking bay in the goat-house to which the animals are taken from their pens.

Part of a well designed and equipped dairy.

The construction of a dairy is similar in principle to that of a milking parlour in that walls and floor should be of washable impervious materials and horizontal surfaces should be kept to a minimum. Benches or tables should be topped with stainless steel or laminate and designed so that the floor beneath can be cleaned easily. Cupboards for storing equipment and materials should be built-in and sealed to walls and ceiling. A sink for washing equipment together with, in larger enterprises, one or more dairy wash tanks for bigger items will be needed. A separate handbasin is essential.

Sufficient electrical power should be available for refrigerators and freezers for storing milk and dairy products. Hot and cold water services are essential for the dairy, although cold water alone will be sufficient for the parlour. Both parlour and dairy require good levels of illumination, and windows should be fitted with removable mesh panels to prevent the ingress of insects.

The other equipment for dairy and parlour will depend upon whether goats are to be hand or machine milked, and the range of products. It could include, for example, cream separator, cheese moulds, cheese and yoghurt vats and so on.

A separate dairy is not necessary where produce is for home consumption only.

Control

Nowhere will the goat exercise its natural inquisitiveness and ingenuity more actively than in trying to circumvent the measures that its owner takes to keep it under control. Since the owner is legally liable for the costs of damage (and road accidents) caused by stock straying, it is not a matter to be dismissed lightly. If management is poor, and goats are kept on overgrazed pasture, they will continuously seek for weak spots in fences in order to escape to lusher pastures.

Fencing

Where sufficient pasture is available to allow goats to graze freely, three types of fencing have been used with varying degrees of success.

Chain link (galvanized, not the plastic-covered variety) is probably the most secure form of fence although its cost is prohibitive. Strained, high-tensile wire fencing has been used to control goats, but is only really successful where the distance between the wires is sufficiently small to discourage stock from escaping. It is not usually suitable for young kids. Concrete posts, well concreted into the ground, are best for strained wire fences but are expensive. Woven stock fencing, of the gauge sold for cattle and pigs, is adequate for goats but for the fact that they tend to climb up the fencing and may, in time, cause it to collapse.

Post and rail fencing is not suitable on its own and lighter gauge nettings (e.g., sheep netting) are better with several electrified wires inside them.

All fences should be at least 1·2 m (4 ft) high. Barbed wire should be avoided.

Hedges and walls

Unlike cattle and sheep, goats are not respecters of 'stock-proof' hedges. The traditional farm hedge, composed principally of hawthorn and blackthorn, will be destroyed rapidly and cannot be relied upon to contain goats for prolonged periods.

Walls of brick or stone 1·4 to 1·5 m (4 ft 6 in to 5 ft) in height are goatproof providing that the surfaces are free of projections that could offer footholds. In cases of doubt, a single electrified wire 900 mm (3 ft) from the ground erected about 300 mm (1 ft) inside the boundary, will strengthen the defences.

Electric fencing

Electric fencing is comparatively cheap to install and to run. It has the additional advantage of being moved easily and therefore can be used to strip graze pasture and forage crops.

A mains or battery powered fencer unit causes electrical pulses to pass through the fencing wires connected to it. Any animal coming into contact with an electrified wire will receive a shock which, whilst it causes no harm, is of sufficient strength to encourage it to avoid the wires in future. Stock meeting electric fencing for the first time should be trained to it by being held against the wire for a moment or two.

Electric fences use little or no power except when the circuit is completed by an animal (or other conductor, such as vegetation) touching the wire and allowing the current to run to earth. It is essential, therefore, to ensure that proper insulators are used, for example, where the wire is attached to posts, and that all vegetation or other materials that could cause short-circuits are kept away from the fence. To neglect this will result in a weakening of the pulse, (and hence a reduction in the efficiency of the fence) and an increase in running costs due to more rapid discharge of batteries (or, with mains units, an unnecessarily high power consumption).

Three wires are needed to contain goats. Heights of 1·2 m, 600 mm and 300 mm (4 ft, 2 ft and 1 ft) above ground level are satisfactory. The lower wire need not be electrified unless young kids are grazing with the herd. Although more expensive, stranded steel wire will last far longer than polythene with fine conducting wires woven into it. Great care must be taken when erecting electric fencing in the vicinity of high voltage overhead power lines: **this is dangerous, and can kill.**

All in all, electric fencing is probably the method of choice for restraining goats when cost-effectiveness and efficiency are taken into account. However, occasional failures will occur and for 'sensitive' locations (such as boundary fences and those protecting orchards and gardens), the effectiveness of the electric fence will be improved if it is erected in front of another barrier (such as a hedge or sheep netting). Electrified netting is available but few goat-keepers have used it with success.

Tethering

Tethering is not a particularly effective method of control. It is time-consuming for the owner, can be detrimental to the health of the goat and, in extreme cases, can result in the animal's death if not conducted with care.

Tethers will need to be moved frequently (say three or four times each day) to give animals access to fresh grazing. (They will not eat grass that has been fouled or trampled). Tethered goats will have to be moved to shelter in bad weather and will have to be checked regularly to ensure that the tethers have not caught up on bushes or other obstructions.

Tethering is not suitable for kids or goats of an active or excitable temperament: neither is it suitable where animals are likely to be scared by sudden noise (e.g, passing traffic, aircraft, etc.) or tormented by dogs or children. It cannot be used on ground that is obstructed (e.g., with trees, bushes, rocks, stumps, etc.) or is steeply sloping. If a tethered goat panics and pulls on its collar, the resulting pressure on a nerve in the neck can result in collapse due to a temporary paralysis of the diaphragm and heart. Recovery is normally spontaneous and rapid, but death can result where the position of the collapsed animal maintains pressure on the neck, for example, where it has been tethered on a slope.

If tethering is to be done, chain must be used (not a rope or any other material that would be likely to kink) one end of which should be clipped into a stout collar. The chain should have at least two swivel links and be attached to the stake in such a way as to rotate freely around it, in order to reduce the chances of it becoming tangled. Stakes should be hammered firmly into the ground and positioned so that other animals are not endangered. It is extremely dangerous practice to tether some animals whilst leaving others to run free in the same field.

A running tether, involving a wire stretched between two stakes and to which the tethering chain is attached, will give the goat a greater area of grazing and may be an improvement over the usual kind.

5 Feeding

Dietary requirements

A goat requires a diet that includes protein, carbohydrates for energy, minerals, vitamins and water. These must be provided in correct balance to keep the animal healthy and productive. Protein is needed for growth, repair of tissue and for production. It is obtained mainly from concentrates and, to a lesser extent from bulk foods. Energy for metabolism is provided by carbohydrates, which are present in both bulk and concentrated foods. Minerals and vitamins are essential for correct functioning of a variety of bodily processes, and are available from many different foodstuffs. Water must be freely available to the goat. Not only is milk composed of a high proportion of water, but water is a major constituent of all body tissues.

The goat as a ruminant

The goat is primarily a browsing animal, depending on high fibre, bulk foods for its nutrition. Because of the high fibre content of the diet, the goat's digestive system needs to be extremely efficient to break down and utilise the cellulose. Along with cattle and sheep, the goat is a ruminant, which means that food is chewed and swallowed into the rumen where the activity of micro-organisms starts to break down the food. The food is then regurgitated into the mouth, chewed again, and swallowed, this time by-passing the rumen and going into the reticulum, then to the omasum and finally to the abomasum, the true stomach. Digestion continues throughout this passage, leaving very little waste material to be excreted.

A goat in full production requires more dietary protein than that generated by microbiological activity in the rumen. Feedstuffs containing proteins which are not attacked by the ruminal flora will pass further down the alimentary tract to be digested there thus providing a further source of available protein. Feeds whose proteins are easily degraded in the rumen include hay, silage and unheated soya bean. The proteins of fresh grass and legumes, dried grass, cooked soya bean meal, linseed cake and fishmeal are subject to a moderate degree of degradation in the rumen. Least degradation occurs with lucerne pellets, dried sainfoin, herring meal, and meat and bone meal, which consequently form the most valuable feeds.

Providing the dietary needs

The dietary needs of the goat, as outlined above, are provided from two sources, bulk foods and concentrates.

In the wild, the goat survives on bulk foods only, but, although the domesticated goat can exist for maintenance on bulk foods, foods of a high protein level must be fed for production. Grain in various forms and high protein oil seed cakes and legumes are used, generally termed 'concentrates'.

Systems of management

How a goat is fed depends very much on how it is kept. A stall-fed goat, which remains in a building, only going out for exercise for a short period each day, will require a lot of greenstuff cut and carried to it, as will a goat spending much of its time in an exercise yard.

If good pasture is available for grazing, or extensive areas of scrub for browsing, a goat will be able to obtain a high proportion of its bulk food for itself on a free range system. As the grazing begins to lose its quality in July-August, extra green food will be required, although goats browsing will find a certain amount to eat well into the winter.

The natural feeding pattern of a ruminant is to feed for a time to fill the rumen, then to lie down to cud for a time and then to continue feeding. To obtain the best results, it is worth feeding goats to this pattern, so that they can make the best use of the food given to them.

Bulk foods

The goat obtains most of its bulk food in the form of grass and herbs, browse, that is, twigs and leaves, and its mainstay for feeding throughout the year, hay.

Hay can be made from grass and herbs on a permanent pasture and is known as meadow hay. If grasses, often with clovers, are sown specifically for growing a hay crop, the hay is known as 'seeds' hay. Excellent hay can be made from lucerne. The best hay for goats, now virtually impossible to obtain, is red clover.

The quality of hay varies enormously, since the time of cutting and the weather are all critical. Good hay, made just as the grasses come up to flowering provides a significant proportion of the bulk food required by a goat, but poor hay, even if palatable, will contribute little to the goats' nutrition. Hay should be offered *ad lib.* with no restrictions in quantity. During the summer, a goat out grazing each day will eat probably only 450 to 900 g (1 to 2 lb) per day, but in the winter may well eat 2·2 to 2·5 kg (5 to 6 lb). A bale, very roughly, weighs 25 kg (56 lb), so quantities required can be estimated.

During the summer months, and in some regions for some of the winter months, goats can forage for food if grazing or browsing is available. Where grazing is not available, it is often possible to find rough patches of land or verges of lanes carrying only occasional traffic. These can be cut and the green food carried to the goats. Branches cut from hedges and trees (with the owner's permission) can give a welcome addition to the diet, and leaves can be dried for feeding in the winter. Ash, elm (where available), elder, sweet chestnut and hazel are all liked. Avoid branches with thorns. Although these may be popular, they could damage the goat, particularly the udder.

There is a wide variety of crops that can be grown on a garden or small field scale, which can provide a substantial part of the goat's diet. Various kales, cabbages and root crops can be fed from midsummer right through the winter, and comfrey, lucerne and maize can be grown for spring and summer feeding. Details of growing these crops are given below. Garden waste such as broad bean and runner bean plants, pea haulm, split cabbages and sprout and broccoli plants can all be utilized, but tomato plants, potato haulm and rhubarb leaves are all poisonous and must not be fed.

The importance of feeding a good mixed diet of high-quality foods cannot be emphasized too strongly. Whether a goat is grazing on free range, or is yarded or stall fed, good quality, well-made, leafy, sweet-smelling hay is the first essential throughout the year. Allow ample greenstuff, but beware of feeding too much of one kind of food. Too many brassicas can lead to a mineral imbalance, as can too much legume feed. Always try to simulate the goat's natural behaviour, which is to take a little of one type of food, then move on to another, and to stop and cud and not overfill herself at one go.

Concentrates

The high protein feed, known as 'concentrates', comes in the form of various grains, with added legumes, such as kibbled beans and soya bean meal, or oil seed-cake, such as linseed flakes. It has been custom for many years to feed processed grains such as flaked maize, rolled or crushed oats and bran. Recent work on the nutrition of sheep has indicated that whole grains such as barley may be better utilized by a ruminant than the processed grain, but although the feeding of whole grain is practised by some goat-keepers, it has not yet become widespread.

Many feedstuff merchants and millers can supply concentrate mixes suitable for cattle and sheep that can be fed to goats. A digestible protein level of 16 to 18% is required for milking stock, and most dairy mixes and nuts have protein levels of this order. Ewe and lamb mixes generally have around 14% digestible protein, which is probably adequate for an average milker with access to good hay and ample green food. Often ewe and lamb mixes are only available seasonally since ewes are not fed concentrates all the year round. Several of the big feedstuff manufacturers and some local mills are now preparing goat mixes, which, although expensive, are highly suitable and save a lot of effort mixing at home.

Home-mixing can be cheaper than buying in prepared feeds and gives the goat-keeper the freedom to alter proportions for different ages and conditions of goats, and to take into account the other food available. In the spring, when the protein level of the grass is high, the protein level of the concentrates can be reduced, until the grasses come up to flower and the feeding value deteriorates rapidly.

Some suitable concentrate mixes are as follows (all in parts by weight):

1. 1 part flaked maize
    1 part bran
    1 part rolled oats
    1 part linseed flakes or soya bean meal

2. 2 parts rolled oats
    1 part kibbled beans
    1 part bran

3. 3 parts whole barley
    1 part linseed flakes

To all of the above mixes, a mineral supplement must be added. Details are given later in this chapter.

Concentrates are fed according to the rates indicated below, usually in two feeds, one in the morning and one in the evening. A very high yielder will probably not take all the concentrates it apparently needs, but by offering them in three feeds a day, most will be taken.

Kid 0-6 months   per day: 225 g (½ lb)
    6-12 months per day: 225-450 g (½-1 lb)

Goatling 1-2 years   per day: 450 g (1 lb)

Milker   2 years +   per day: 450 g (1 lb) + 1·8 kg (4 lb)
per gallon of milk produced.

Feeding rates for males are included in Chapter 6.

Although the quantities quoted are what the average goat requires, a good stockman will always know when a particular goat needs a little more, or is satisfied with a little less. However, never be tempted to try to improve an animal in poor condition by pushing extra concentrates into it. It is much sounder feeding practice to let it have the best hay available and a variety of green foods.

Concentrates can be fed in clean bowls or troughs, or in buckets attached to the wall or door by a metal ring. Details of such attachments are given in Chapter 4. Bowls should always be removed after the feed is finished to avoid them being jumped on. Troughs are usually too unwieldy to remove after each feed, and tend to get fouled easily. Consequently, care must be taken to keep them clean.

The most satisfactory time to feed concentrates is after milking. Some people feed them during milking, which seems somewhat illogical since a goat's natural reaction is to cud during 'let down' of milk, and most goats will stand and ruminate quietly whilst being milked.

A very popular feed which lies between concentrates and bulk foods is sugar-beet pulp. Soaked in hot water (and drained if surplus water is present), it is fed warm. It can be dried off with bran, particularly if bran does not form part of the normal concentrate ration. Sugarbeet pulp makes an excellent extra feed in the winter or on wet days in the summer when the animals cannot go out to graze. Allow 225 g (½ lb) dried pulp or nuts per goat.

Water

Water should be offered in buckets, either plastic or galvanized. Water should be available for all age groups of goats, preferably warm, and may be lightly salted (1 dessertspoon to 9 litre [2 gal] water). Offer fresh water at least twice a day. Some goats prefer water to be left with them. Ensure goats out grazing have access to clean water, or are offered it at intervals. They should not be permitted to drink from ponds or stagnant water.

Minerals

Goats have a very high mineral requirement compared with cattle and sheep. Part of their requirement is supplied by their desire to graze deep-rooted herbs rather than shallow rooted grasses. The herbs bring minerals to the surface, making them available to the stock. Goats should always be allowed free access to a mineral lick, and, as already mentioned, a suitable mineral supplement should be added to their diet. Ones recommended for goats are available, and sheep mixes are also suitable. Dairy cattle supplements are not recommended because of their copper content. However, they have been fed by goat-keepers for many years with no apparent adverse effects. Horse and pig supplements are definitely not suitable owing to their particularly high copper levels. Always feed minerals at the recommended rate. Too much or too little can cause serious mineral imbalance, leading to metabolic disorders. Free access to a mineral lick, preferably cobaltized, should be allowed. Mineral licks have a high proportion of salt (sodium chloride). A goat secretes 1·5 g of salt in every litre of milk and so requires a very high intake. The following minerals are those of most importance to the goat:

Calcium and phosphorus

Both minerals are essential to maintain bone structure and milk production. The correct balance is essential and a varied and mixed diet usually supplies the correct levels. Care should be taken when adding seaweed meal to the diet, and in the feeding of legumes and sugarbeet pulp, as these are all high in calcium. High phosphorus foods include cereals, oil-cakes and young grass.

Cobalt

The cobalt requirement of a goat is particularly high compared with cattle and sheep, and lack of it results in poor condition and a 'goaty' taste to the milk. Cobalt provides the material from which Vitamin B_{12} is synthesized, which is essential to the animal's health. Parasitic roundworms in the gut will rob the goat of Vitamin B_{12}. Consequently, worm-laden goats usually have 'goaty' tasting milk. Worming should be accompanied by treatment with cobalt salt made as follows:

Dissolve 25 g (1 oz) cobalt sulphate in 0·25 litre (½ pt) water. Wet 2·5 kg salt (6 lb) with the solution and dry in a low oven. Feed 25 g (1 oz) in the concentrate feed every day for a week.

This can be given to a newly-kidded goat during the first week after kidding, when she is extremely susceptible to worm infestation, and consequent cobalt depletion.

Copper

Copper deficiency is prevalent in some areas of Britain, particularly fen areas and some grassland. The deficiency causes a condition known as 'swayback'. In 'swayback' areas, small quantities of copper sulphate must be fed to prevent the condition, which takes the form of paralysis of young stock. Deficiency also causes poor condition, infertility, staring coarse coat and scouring in adult stock. Care must be taken when adding copper sulphate to the goats' diet as it is highly toxic. Veterinary advice on suitable levels should be sought.

Copper poisoning can be brought about by goats being fed unsuitable mineral preparations such as those intended for horses, and by feeding commercial pig rations to goats.

Magnesium

Magnesium deficiency can occur in animals feeding on grass that is growing fast in the spring and has a low magnesium level. The deficiency causes lactation tetany. A mixed diet including deep-rooted herbs will help to prevent this. Cattle grazing lush grass in the spring are often given magnesium supplements in their concentrates to prevent 'grass staggers'. Since goats rarely eat only grass, the condition is not as common as in cattle.

Iodine

As with copper, some localities in Britain have a shortage of iodine in the soil. This affects the function of the thyroid gland causing low metabolic rate, dry skin and poor coat condition, and stillbirths. A serious deficiency results in goitre. A kid in the stage of fast growth may develop a goitre-like swelling under the throat. This is temporary and usually corrects itself as the growth rate slows, but iodized salt may be added to the drinking water to help this condition.

Unlike copper deficiency, addition of the mineral to the diet is not successful, and treatment of pasture with Chilean nitrate or seaweed fertilizers is more satisfactory.

Vitamins

Vitamin A can be manufactured by the goat from the precursor, carotene, which is present in fresh grass, green foods such as cabbage and kale, and also carrots, roots and maize. As long as fresh feedstuffs and well-cured hay are available throughout the year, deficiencies are unlikely, particularly since the vitamin is stored in the liver and can be released when the dietary levels are inadequate. Vitamin A is essential for maintenance of the skin and mucous membranes, and aids resistance to infection. Fertility can also be affected by Vitamin A deficiency.

Vitamin B is synthesized by ruminal bacteria, so deficiencies are unlikely unless the diet is particularly low in fibre (which would reduce the bacterial population) or the goat is carrying a large worm burden. If deficiencies do occur, apart from proprietary vitamin supplements, sources of Vitamin B are bran, wheatgerm and yeast extracts.

Vitamin D is necessary for calcium and phosphorus absorption and metabolism. It is synthesized in the skin on exposure to sunshine and skyshine, although the vitamin is present in small quantities in herbage and well-made hay. Deficiencies will cause rickets in growing animals and osteomalacia, a condition producing brittle and weak bones, in adults.

Vitamin E is widely available in livestock feeds but deficiencies can occur, particularly where crops and soils are deficient in the trace element selenium or where cod liver oil is regularly administered to stock. Vitamin E is required for normal reproduction and deficiencies can cause muscular wasting in ruminants.

Pasture and pasture management

Goats are not grazers by nature, but are browsers. However, they will do very well on a good mixed ley containing various herbs as well as a variety of grasses. Permanent pasture that is kept in good condition is ideal, but poor weedy patches will not contribute much to a goat's diet. If the chance comes to re-seed a field, the best choice is a herbal ley containing several grass species, and a mixture of herbs to include white and red clovers, sheeps parsley, burnet, chicory, yarrow, vetches, plantain and dandelions.

Goats like to move around in between periods of grazing and are not the best of stock for strip grazing. To control them with an electric fence, a wire would need to be run at low level, and would consequently touch the vegetation and short out. The best use of pasture is made by dividing it into small paddocks, allowing one to be grazed and then moving the goats on to a fresh paddock. The paddocks can be divided by electric fencing, but a determined goat will get through electric fencing unless it is backed by a more solid fence.

To keep the pasture in good condition, the grasses and herbs should not be allowed to go to seed, since they then become tough and stalky and lose their palatability and food value. If the goats are not keeping it under control, the grassland should be topped to encourage new growth.

Pasture should be rested every few weeks to recover its growth and prevent a worm burden building up. Roundworms parasitic in goats' guts spend part of their life cycle on grass. By resting the grass for a time, the life cycle of the worms can be interrupted.

Pasture should always be treated as a crop, not as something that will grow without attention. Liming can improve a paddock that is getting sour, and a dressing of fertilizer or well-made compost or farm-yard manure improves the fertility considerably. A soil analysis will show the precise fertilizer and lime requirements. Analyses are undertaken by the Ministry of Agriculture or major fertilizer manufacturers.

Growing crops

With careful management and cultivation of a garden plot, allotment or part of a field, a good proportion of a goat's green food requirements throughout the year can be grown.

Comfrey

Giant Russian Comfrey will give early green food, starting to grow as soon as late March to early April in a mild spring. Goats like the leaves when they are young, but find them less palatable when the plants run to flower. Flower heads should be cut off and new leaf growth encouraged. The roots can be purchased, or obtained from other goat-keepers. Once a bed is established, it will continue to grow for many years. The plants should be set about 450 mm (1 ft 6 in) apart and mulched liberally with well-rotted muck each winter. Propagate by digging up the roots, splitting them and re-planting in the winter.

Lucerne

Lucerne is a leguminous plant, much liked by goats. Once established, it will stand for five to seven years, and can be cut several times each summer. The ground must be free of weeds before sowing in April or late July, and should be neutral or alkaline and have a high level of phosphate. An inoculant will probably be needed to encourage germination of the seed. This can be obtained from the seedsman.

Lucerne is deep-rooted and consequently drought-resistant and makes excellent green feeding. If sufficient can be grown, it will make very palatable, high-quality hay.

Oats and tares

Tares or vetches are low-growing, sprawling plants. If grown with a cereal such as oats, they will use the cereal stems as supports, thus making cutting of the crop easier. An agricultural seedsman will supply a seed mix to which field beans are sometimes added. This should be sown on limed land with an ample phosphate and potash level between February and April, for cutting during the summer.

Chicory

Whitloof variety of chicory can be grown as a row crop to give several cuts of the dandelion-like leaves which are very much enjoyed by goats.

The crop will stand for several years, but tends to run up to flower after the first season and is not so palatable. The seed grows well on most soils, and, being deep rooted, the plants are drought resistant.

Maize

Several varieties of fodder maize are now available. Maize can be grown to cut for feeding in the late summer and early autumn. The plants are tall (1 m–1·2 m [3-4 ft]) and the whole plant is cut when the cobs are formed. The plants are not frost hardy, and must be fed before the winter frosts start. It grows best on well-manured, light soils, but does not do well on heavy soils. The seed is sown in May at a depth of 60 mm (2½ in) and if only a small plot is sown, it should be sown in a square patch rather than long rows, to aid fertilization. Rooks love the seed and the seedbed must be protected from them.

Cabbages

A succession of different varieties of cabbage can be grown on firm, well-limed soil to provide feed throughout the year. The dark green varieties are better feed value than the white ones, but both are well liked.

Kale

Kale, again, thrives on well-limed land and several varieties can be grown for a succession of feeding from early autumn until spring. Marrow stem is the first variety, but is not very frost hardy so is usually sown in late April to early May for cutting before Christmas. Maris Kestrel and Thousand Head kale are both frost hardy and can be kept to feed after Christmas until they start running up to flower in late March-April. Kales are sown in drills 450 mm (1 ft 6 in) apart and thinned to about 300 mm (1 ft) apart in the row. Better growth is achieved on rich soils, but kale will do well on poor soils and can be grown in most regions of the British Isles.

Roots

Various root crops such as swedes, fodder beet, mangolds and sugar beet can be grown to feed, either chopped up or sliced in half and left for the goats to gnaw at. Roots, except perhaps swedes, must never be fed to male goats as they can cause urinary calculi which block the urethra. The wilted tops from the roots can be fed to female stock. Roots all do well on fertile soil and whilst doing best on light lands, grow very satisfactorily on heavy land. Late April to early May is the normal time for sowing to allow a long growing period, although swedes can be sown up to early June. It is usual practice to lift mangolds in October and store them until after Christmas. Fodder beet does not require lifting as it is more frost hardy and can be pulled or dug as required. Sugar beet can be treated in the same way as mangolds. Swedes are best lifted and stored although they will stand the winter in milder regions.

Carrots are an excellent root to feed during the winter, but the yield from a garden crop is probably not worth the space they take. However, anyone living near to carrot washing plants can usually obtain bags of reject carrots cheaply.

Poisonous plants

Goats eat such a wide variety of plants that many which may be considered moderately poisonous can be eaten in small quantities with no apparent harm. It is advisable to give a goat a feed of hay before putting it out to browse or graze, so that it is not likely to gorge itself on one particular plant, and to check paddocks regularly for any poisonous plants which may have grown. Some plants are relatively innocuous at certain stages of their growth and become more poisonous at other times, particularly when they are fruiting. Parts of some plants may be poisonous, whereas other parts of the same plant may be relatively harmless.

Probably the safest approach with garden plants and evergreens is to consider them all as poisonous (see p. 92).

6 The male goat

Although an artificial insemination service for goats is not available in Britain, there is no real need for the household goats milk producer to keep his or her own stud male. In most parts of the country there are pedigree bucks from high milk producing lines standing at stud. Stud fees are not high (compared to those for other livestock). Although it is not possible to be precise, it is probably not economic to keep a stud buck of one's own unless one has between, say, six and ten does to be mated each year. Apart from the costs of feeding and husbandry, a male goat will require housing (of more substantial construction than that for females) and will add to the owner's work load. In addition, bucks are not the most pleasant of animals to handle due to the strong and clinging odour they produce (as a sexual attractant), their habit of ejaculating and spraying urine on their surroundings, and the fact that, like other male farm stock, they are strong and can be aggressive. The advantages of having a male goat on the premises are that matings can be achieved without major upset to the work of the house and small-holding, and matings early or late in the breeding season (or out of season) can be made with more success.

Choice of buck

Whether the question is of selecting a buck to purchase or deciding which one to take does to, there are several points to be taken into account.

The most important feature is the ability of the buck to transmit deep milking qualities to his daughters. With a relatively small goat population, there is not the same scope for progeny testing as in dairy cattle. However, a step in the right direction is the British Goat Society's Sire of Merit award. This award (designated by the letters SM before the buck's name) is given to male goats who have sired five or more daughters that have achieved above minimum national standards either for milk yield in an officially recorded lactation or for milk yield and butterfat given in a twenty-four hour milking competition at a show recognized by the British Goat Society.

Due to the comparatively long period for the buck's daughters to reach maturity, the SM award is often made late in life or posthumously! In most cases, one will be forced to consider the milking qualities of the buck's ancestors rather than those of the progeny. A male whose dam and sire's dam have each qualified for either the * or Q* in a show milking competition is himself entitled to the Dagger (†) prefix. One whose dam and sire's dam have each been officially milk recorded is entitled to the Section Mark (§) which carries with it an index to the recorded yields. For example, § 310/275 † Polyglot Palmer would be a buck whose dam had given not less than 3,100 kg of milk in an officially recorded lactation period of one year and who had gained the * or Q* award, and whose sire's dam had given not less than 2,750 kg and who had also been awarded the * or Q*. The milk yields of any of the buck's full sisters can also give useful pointers to his potential. In general, the conformation and appearance of the buck is of minor, if any, importance providing, of course, that he is in sound health and free from hereditary faults, such as double teats.

Bucks can be purchased at any age although it is often more successful to buy a kid at four days of age and to rear him. Home-reared bucks are often more easily managed than those bought in when partly or fully grown. An adult male will be capable, on average, of serving two or three does each day, providing he has an adequate break between them. A male kid is usually restricted to two or three services a week in his first season.

Rearing

Buck kids are reared in essentially the same way as females. Due to their faster rate of growth they are often given up to 3 litre (5 pt) of milk each day. Some goat-keepers feed milk for a longer period than is the case with females, often giving a bottleful every day through the kid's first winter. Concentrates should be introduced as soon as possible and gradually increased so that 450 g (1 lb) a day is consumed by six months of age. Bucks are capable of giving effective services from about twelve weeks old and should be separated from females by then.

Feeding

The amount of concentrates fed after six months of age should be gradually increased to 900 g (2 lb) per day, in line with the buck's growth. Bucklings and adult males will require about 900 g (2 lb) per day through the summer (providing of course, that they have access to adequate green food) and up to 1·8 kg (4 lb) daily through the breeding season. The onset of the breeding season often results in a loss of appetite and it is therefore a good idea to begin increasing the concentrate ration from late June or early July onwards. The object is for the male to start the breeding season in good, but not overfat, condition.

A diet of foods rich in calcium can cause infertility in bucks due to the formation of urinary calculi, resulting in blockage of the urethra. Care should be taken to avoid root crops (except carrots and, possibly, swedes), legumes and legume hays (clover, lucerne) and to restrict the amounts of other calcium-rich foods, such as kale. In hard-water areas it would be good practice to use rain water for drinking. A copious supply of clean water will encourage the male to drink sufficient amounts to reduce the chance of calculi forming.

Grazing is beneficial for bucks, providing them with an opportunity for exercise as well as fresh green food, but a good standard of fencing is required. It is also an advantage if the buck can have some company whilst at grass: they will happily share their pastures with calves or sheep. If insufficient land is available to keep the male in a separate field from the milking herd, he can be put out to graze, in the spring and summer at least, after the herd has been brought in for milking.

Housing

To avoid unplanned matings and the problems that result from his odour, the buck should always be housed separately from the rest of the herd. However, to reduce boredom (and aggression) it is an advantage if his housing is positioned so that he can see the other goats. If he can be penned alongside other stock, so much the better.

Male goat housing must be of sturdy construction. As a general guide, an adult buck can be regarded as equivalent to a bullock in strength. Providing that the housing gives adequate shelter, there is no need for it to be enclosed on all sides. About 6 sq m (60 sq ft) floor area should be allowed for a shed or pen for a single buck. The roof should allow a minimum internal clearance of 2 m (7 ft). It is an advantage if the pen opens directly on to a concrete exercise and service yard, and if feeding and watering can be done from outside the pen.

Handling

The principal difficulties in handling bucks result from their strength and smell. Like other male farm stock (bulls, stallions, etc.), they can be aggressive and should be handled calmly, firmly and with respect. Aggression can be minimized by adequate exercise and company.

7 Breeding

Breeding cycle and season

Goats in the temperate regions of the world breed only during the winter. Like sheep, they generally come into heat during September and October. Thereafter, the period of heat (or season) repeats approximately every twenty-one days, unless they are successfully mated, through to late February or March.

The breeding or oestrus cycle is controlled by hormones. Onset of the breeding season is triggered by decreasing day length. As the breeding cycle starts, a noticeable drop in yield occurs in milkers. No amount of extra feeding will prevent this drop.

Signs of heat include persistent bleating, tail wagging, mucous discharge from the vulva, poor appetite, unusually awkward behaviour and restlessness. A goat may show one, some or all of these signs, usually mildly at the beginning and end of the breeding season, but very strongly at the height of the season. The period of heat lasts from a few hours up to three days. It is advisable that the goat is mated on the first or second day.

Age to breed

In Britain, it is usual practice for a goat to be mated when she is approximately eighteen months old, to kid for the first time at around two years. A well-grown kid may be mated when she is nine or ten months old. Kids born in December and January can be mated under a year of age to kid in the spring, thereby avoiding the unproductive goatling stage.

Mating

Prior to the breeding season, it is advisable to plan which goats to 'run through' (i.e., to leave unmated as winter milk producers) and which goats to breed. Stud males should be carefully selected. The British Goat Society and many of its Affiliated Societies publish stud goat lists in the late summer, advertising males of different breeds. The distinguishing signs that show the milk producing history of their immediate female ancestors are explained in Chapter 6. The owner of the stud goat should always be contacted in advance to ensure that the buck will be available for use.

Some does will need time to settle on arrival at the stud goat owner's premises. Even if the signs of heat have not been strong, a doe in season will look alert on smelling a male and hold her tail erect or wag it frantically. The owner usually holds the doe by her collar while the buck is allowed to sniff round her. If she is receptive, she will stand still with her tail up. The buck will mount her and mate fairly quickly. Mating is brief, after which the animals are separated. Many stud goat owners allow a second mating about ten minutes after the first, although whether this is of any benefit is doubtful. The owner of a pedigree buck will issue a stud certificate for the mating, which must be sent to the British Goat Society when any kids are registered.

Pregnancy

Pregnancy lasts for about five months (150 $\pm$ 3 days) although kids may be born a week or more either side of 150 days. An in-kid goatling will require very little extra care during pregnancy. High-quality roughage should be fed to ensure that she keeps in condition, but it is unlikely that extra concentrate feeding will be necessary. An in-kid milker may well drop rapidly in milk and go dry, but if not, she should be dried off by eight weeks before kidding. To dry her off, the goat is not milked right out, and as the yield drops, milking can be reduced to once a day. It is helpful to reduce the amount of green stuff fed during the drying-off period, but this can be increased again once the goat is dry. Many goat-keepers like to 'steam up' their goats before kidding by steadily increasing the amount of concentrates fed during the last eight weeks of pregnancy until the goat is eating about 1·8 kg (4 lb) a day just before she is due to kid. Some people view this as excessive and lay the blame for some metabolic diseases on heavy concentrate feeding at this time. The alternative is to continue a maintenance ration of 450 g (1 lb) per day of concentrates plus ample high-quality hay and green food.

Pregnancy testing based on an assay of a milk sample, is now available. The milk sample, collected twenty-three days after mating, is sent to either the Milk Marketing Board laboratory or a private laboratory. The local veterinary surgeon will be able to provide details.

The developing kids are well-protected inside the goat. As long as care is taken when the goat goes through doorways, and direct blows are avoided, little extra care is needed. The doe's feet should be trimmed by about the third month of the pregnancy. They may then be left until after kidding. It is advisable to vaccinate against enterotoxaemia and tetanus during the last two to three weeks of the pregnancy. This will also protect the kids after they are born, since the immunity is passed on via the milk.

Breeding problems

Sometimes a goat will not come into season, or comes into season irregularly and will not hold to service (i.e., conceive). Hormone and prostaglandin preparations can be administered by veterinary surgeons to try to establish a regular oestrus cycle. There are several causes of infertility in male and female goats. Mineral deficiencies, particularly of iodine and copper, can lower fertility in both sexes, as can obesity.

Hermaphroditism may not become evident until the goat is adult. A hermaphrodite is always infertile. When once the abnormal formation of the vulva is identified, and masculine secondary sexual characteristics, such as thickening of the neck and coarsening of the coat, appear, any hope of being able to breed with the animal is in vain.

A goat may develop cystic ovaries at any stage of its life. The cause is unknown. The doe will come into season repeatedly, very often at more frequent intervals than the normal three weeks and for prolonged periods, and will not hold to service. Prostaglandin therapy has been found to be successful.

'Cloudburst' is a fairly common problem amongst goats, though is rarely heard of in other ruminants. It is essentially a phantom pregnancy, now thought to be a form of hydrometritis, which can occur spontaneously or after mating. The goat goes through an apparently normal pregnancy and at term produces large volumes of fluid with no placenta and no kids. The doe may search for her missing kids. Blame for cloudbursts is often put onto the practice of running goats through a two year lactation. The cloudburst then occurs after the second year, leaving a dry or low yielding goat for a further year. However, cloudbursts have been recorded in goatlings and after one year's lactation. Some families of goats seem to be more prone to them than others. Most goats will breed again normally after a cloudburst.

If a doe is in good health and is cycling normally but not holding to service, the male's fertility should be checked. A veterinary surgeon can arrange for a sperm count to be carried out. It is worth checking whether the buck has served other goats successfully. Assuming the buck is healthy and in good condition, there is little that can be done to cure infertility. The use of certain foods can result in the formation of urinary calculi and subsequent impairment of the buck's fertility.

Kidding

As kidding time approaches, the doe should be given an individual pen if she is not already housed in one. It should be clean and freshly strawed. It is preferable for her to have a few days to get used to her new surroundings before kidding.

Several days (or even two to three weeks) before kidding, the goat will begin to 'bag up', that is, her udder will increase in size before the next lactation. Milk, which feels sticky like colostrum, may be present. Only if the udder becomes very full and tight should the goat be milked before she is due to kid. However, easing the milk before kidding will not greatly affect the quality of colostrum, since the vital antibodies will still be present in adequate amounts. Any milk drawn from the udder before kidding can be stored in the refrigerator or freezer and subsequently fed to the kids. Colostrum is the milk produced up to the first few days after kidding. It is essential that kids have it during the first six hours of life, and preferably for the first four days. Colostrum contains antibodies, which confer immunity to disease on the kid. It also has high levels of Vitamin A, needed by the kid shortly after birth, and laxative properties which aid removal of meconium, the black sticky gut contents. A kid that does not receive colostrum is unlikely to survive. Colostrum can be frozen and a useful safeguard is to freeze some from the first goat to kid each year. Should a doe die at kidding, the colostrum can be thawed and fed to the kids. (It should not be over heated, but can be warmed in a water bath).

A few days before kidding, the doe's backbone between the point of the hip and the tail will loosen, making an arch. The vulva will look 'puffy'. One of the first signs that kidding is commencing is that the goat will cease to eat. (Food and water should be removed at this stage. It is advisable not to leave water with the goat for the last few days. Should she kid suddenly a kid may be dropped in the water bucket and drown). The goat will become restless, pawing at her bedding. If she is out grazing with the herd, she will go off on her own to a quiet place. A clear mucous discharge may appear and then, within hours, the goat will be in labour. Finally, she will strain strongly to push the kids out.

The normal presentation of a kid is with its forelegs outstretched and appearing first with the head lying on them. The kid may be lying in the intact water bag or this may have broken. It is very common for the doe to bellow as the kid is born. This may be somewhat un-nerving to the goat's owner. When once the head and shoulders are born, the hind part of the kid follows easily. If the doe is standing to kid, the navel cord attaching the kid to the placenta will generally break as the kid falls to the floor. If the doe is lying, the cord usually breaks as she gets up to inspect and clean up the kid. As soon as the kid is born, the owner or attendant should wipe over the kid's nose, remove any mucus from its mouth and check that it is breathing satisfactorily. After this, the doe will lick the kid to clean it up. If the goat is a first kidder, or the kidding has been difficult, she may not want to lick the kid, in which case the attendant must rub it with a towel or soft hay to prevent it chilling. The kid should then be put into a corner of the pen or into a tea chest lined with hay, placed where the doe can still nuzzle the kid. Further kids are likely to appear up to half an hour after the first one and it is not unusual for subsequent kids to be presented in the breech position (hind legs or rump first). As long as the kid is lying straight and not twisted, the doe should be able to deliver a second or third kid in this position without assistance, although she is unlikely to be able to deliver a breech position first kid unaided.

When all the kids are born the placenta or afterbirth follows, usually within a few hours. This is seen as the fleshy mass hanging from the goat's vulva. Under no circumstances should any attempt be made to pull this, as it could cause internal haemorrhage. If it has not come away within twelve hours, contact the veterinary surgeon who will remove it manually. The goat will probably eat the afterbirth if she gets the chance. Whilst it does no harm to the goat, it does not appear to do any good either and is probably better removed.

Abnormal kiddings

Once the goat is straining hard every few minutes, the kid should appear within approximately a quarter of an hour. If not, the attendant should suspect that something is wrong. The attendant should either phone the veterinary surgeon immediately, or, if confident of carrying out an internal examination, prepare by washing hands and forearms, trimming and scrubbing the nails and wiping lambing oils or gel over the hands.

The hand is gently inserted into the vagina. If it is impossible to get more than a couple of fingers in, the cervix has not dilated and the condition may be that known as 'ring womb'. A veterinary surgeon should be called, who will either open the cervix manually, or carry out a Caesarean section. Assuming that the hand can get into the doe's uterus, it should be possible for the kid to be felt. If the kid is lying forwards, one leg may be folded, or the head may be turned back, preventing it from being born. The attendant gently manipulates the kid into the correct position, pushing it back into the space of the womb if necessary, and then with each uterine contraction, assists the doe by holding the kid's forelegs and pulling gently and firmly. If the

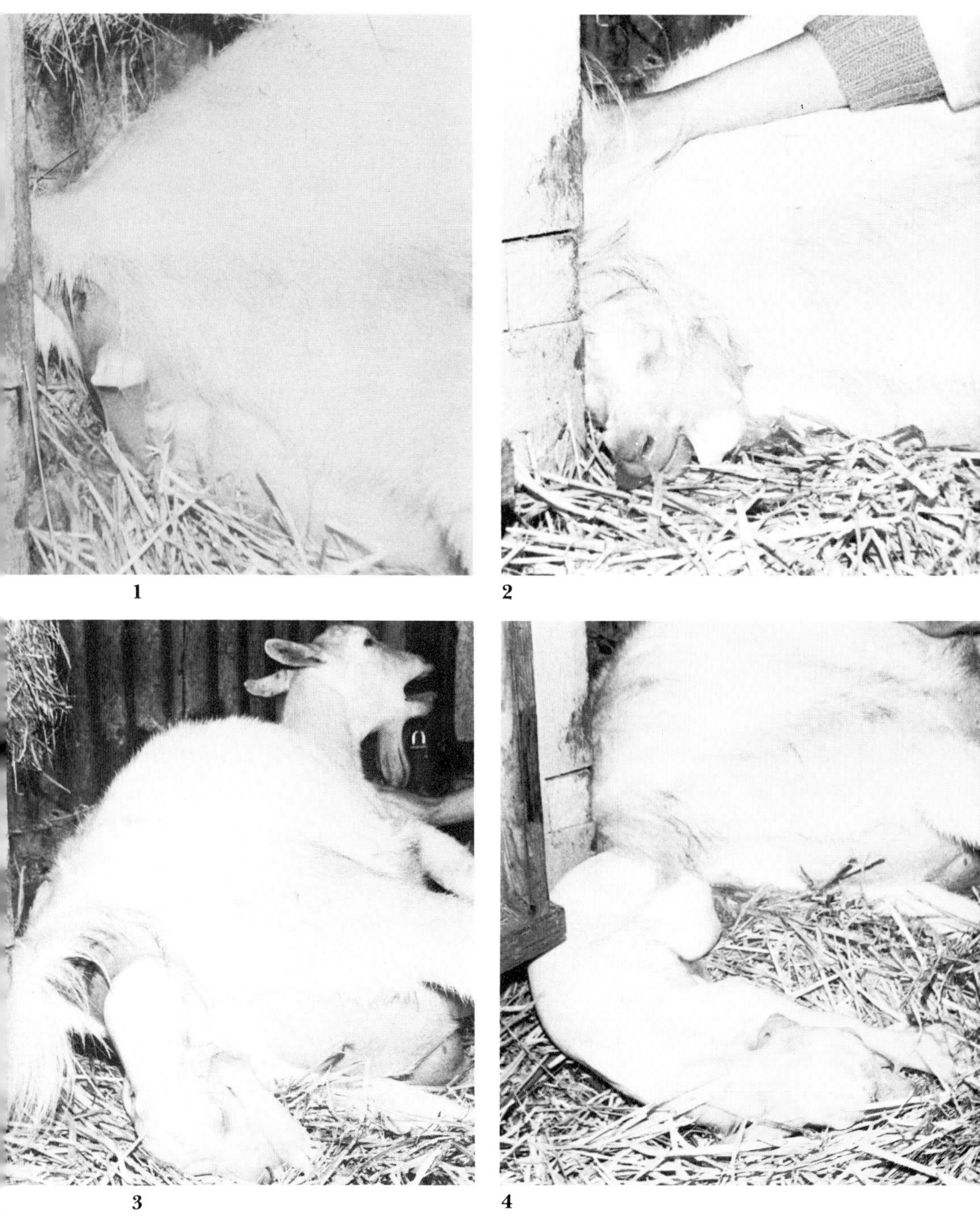

1

2

3

4

5

6

The birth of kids

1. The kid's front feet are just visible. **2.** In a normal presentation, the head and forelegs are delivered together. **3-4.** The goat strains to deliver the rest of the body. **5.** The dam licks and nuzzles the newborn kid. **6.** A second kid is delivered, usually more easily than the first (and often, though not in this case, in a breach position). **7.** The dam cleans the second kid.

7

kid is lying in the breech position (with hind legs or rump first), it can be delivered, usually quite easily, with assistance, again exerting a gentle pull with each uterine contraction. It may be that two kids are coming together, which will seem like a jumble of legs and heads. If the attendant can sort them out by pushing one kid back and pulling one forward, this will help but he should never hesitate to call the vet.

After any manual investigation, the goat should receive antibiotic treatment, either pessary or injection, to prevent infection. After an assisted kidding, the doe is more likely both to retain her afterbirth and to be a poor mother, perhaps not cleaning up the kids or letting them suckle. In this case, it is most important to ensure that the kids receive colostrum from a bottle, and that they are not in danger of being trampled or butted through the doe's fear of them. Often within a few hours the doe will recover and mother the kids quite normally, but sometimes she rejects them totally and hand-rearing is the only answer.

Care of the new born kid

Once the kid is born, the mother will want to lick it to clean it up. This both dries the kid and prevents it from chilling. The umbilical cord breaks fairly quickly, and no attempt should be made to cut it, since it breaks at a point of natural weakness, and any interference could cause a haemorrhage. The kid is left with the doe until she shows signs of straining again, when it can be put into a hay-lined box, within sight of the doe, or laid in a safe corner of the pen.

When all the kids are born, they can be taught to suck from the doe straight away by placing them near the udder and gently putting the teat into their mouths. It is a natural reflex for them to suck, and it is not difficult to get them started. Pushing them on the back of the head should be avoided as they will push backwards from this and struggle. It can help with weaning them on to a bottle at four days of age if they are given their first drink from a bottle. Colostrum is milked from the doe into a warmed jug, and poured into a warmed bottle, and a teat fitted. Again, the kids suck easily, and can later be encouraged to drink from their mother. If the doe has large teats, these can be a problem for the kids to drink from, as can those where the goat has a low-slung udder. If the kids cannot manage to suck from the teats, they can be left with their mother for four days, but must be bottle fed, four or five times a day.

As soon as possible after birth (say within fifteen minutes) the kids' navel cords should be treated to prevent infection entering the abdomen. The cord and navel are sprayed with Terramycin aerosol spray (foot rot spray) or dipped in an eggcup of tincture of iodine.

Most kids are able to stand within an hour or two of birth, although Anglo-Nubian kids can take a day or two to become strong and lively.

After a prolonged, difficult birth, a kid may be chilled and will require warming up. An infra-red lamp is ideal, suspended approximately 0·5 m (1 ft 6 in) above the pen floor. Alternatively, the kid should be brought in to lie by a fire or stove until warm, and can benefit from being wrapped in an old jumper or piece of blanket when put back with its mother. A cold kid will not suck, so only small amounts of colostrum should be offered at a time until its desire to suck returns.

The kids must be inspected to ascertain their sex, to ensure that there are no signs of hermaphroditism (seen as an enlarged vulva on a female kid, with a protuberance from just inside the vulva), and that no supernumary teats are present in either sex. Kids can be checked for horn buds at this stage, but if it cannot be determined whether the kids will be horned or not, another check a few days later will probably be more successful.

Unwanted kids and any kids with supernumary teats or other major faults should be put down at this stage, or reared for meat.

Care of the dam after kidding

Immediately after kidding, the dam will be very preoccupied with the kids, and will continue to nuzzle and lick them. Some clean straw laid over the wet bedding will help to make her comfortable, and she can be mucked out later when she is more settled. The first thing the mother requires is a warm drink. Salt (one dessertspoonful in 9 litre [2 gal] of water) can be added, and a tablespoon of molasses dissolved in the water is also appreciated.

A bran mash should also be prepared now. Approximately 900 g (2 lb) of dry bran is put in a bucket, and boiling water added to make a stiff paste. Again, a tablespoon of molasses or black treacle can make this more appetizing. The bucket is covered with a cloth and allowed to cool, and offered to the goat whilst still warm. Fresh hay should also be offered.

When the afterbirth has come away, the goat can be washed with warm water round her tail and udder to remove any blood and mucus. After a few days, the doe will start to discharge a brick red mucous, and this persists for four to six weeks. This is perfectly normal, unless it becomes foul smelling, in which case an internal infection should be suspected and veterinary advice sought.

During the week after kidding, bran mashes should be offered for the first two or three feeds, gradually making them drier until the bran is

The mother and kids on the day following the birth.

fed dry with a little flaked maize and rolled oats added. By the end of the week, the protein portion of the feed (oil-cake, beans, etc.) can be re-introduced, and the amount gradually increased as the goat's milk yield starts to rise. It is important not to push too much concentrate feed into the goat at this stage, since this can force up the milk yield, thereby increasing the risk of milk fever, and her digestion can be easily upset. Ensure that adequate minerals are available in the form of powdered minerals in the diet, and a salt lick. Cobalt salt (see Chapter 5) can be added to the concentrate feed during the first week after kidding. One dessertspoonful is added to each feed. This helps to counteract loss of condition caused by worm infestation. Roundworms in the gut multiply rapidly shortly after the goat has kidded, and dosing with a proprietary wormer is recommended during the week after kidding to counter this rise.

When the kids are four days old, the doe can start going out to graze again, but if the grass is lush, this should only be for short periods at first. For goats kidding before the end of March, it is unlikely that they would be going out anyway. They can be offered their usual green food and roots, keeping the amounts moderate at first. Good quality hay should be fed *ad lib.* after kidding, since the goat requires enormous amounts of energy at the beginning of a lactation, and the faster she uses up her reserves, the poorer will be her lactation.

8 Rearing kids

Disbudding

Goats are either naturally polled or born with horn buds, which grow rapidly after birth to develop into horns. Whilst horns might look magnificent they are dangerous to goats and goat-keeper alike. A horned goat will tend to bully polled or disbudded goats. Even a docile horned goat, just by a quick turn of the head, can hurt a person or another goat quite badly. For this reason, it is normal practice to disbud kids.

Disbudding may only be carried out by a veterinary surgeon and must be performed under anaesthetic. It is done with a hot iron, either an electric calf iron (although this is, in fact, too small for the horn base of a goat), or a specially designed goat iron. The hair is clipped away from round the horn buds, the kid anaesthetized and the iron (previously heated to a cherry red colour) is applied to the horn bud for six seconds. The burnt area is then smoothed over with the side of the iron. The process is then repeated on the second bud. The horn base of a male is larger than that of a female, and more should be burnt away in front of the horn bud to prevent the growth of unsightly scurs. Disbudding should be carried out at around four days of age, when the buds can be easily felt, but before they have grown into horns.

Earmarking

All goats to be registered with the British Goat Society in the Herd Book, Foundation Book, Golden Guernsey Register or English Guernsey Section must be marked with a tattoo in the right ear. Many Affiliated Societies will earmark goats for members. Individuals can apply to the British Goat Society for their own letters. The earmark consists of two letters, followed by a number, followed by a year letter. A kid earmarked JJ 2 R is a kid earmarked by the authors, the second in the year to be marked, in 1980. The marking is usually done with a spring-release cattle tattoo marker, and antiseptic coloured paste rubbed into the mark which lasts throughout the animal's life.

Although goats may be earmarked at any age, it is not usual practice to do this until they are at least four weeks old. If they can be left longer than this, so much the better since the ear will be larger, making the job easier and distortion of the mark less likely as the ear grows.

At the time that the animals ear is marked, a certificate is punched with the same letters. This certificate must be forwarded to the British Goat Society with the appropriate forms in order to register the kid.

The purpose of earmarking is twofold. It provides positive identification of the animal for entry in the Herd Book, and, in the case of lost or stolen animals, the goat can be traced more easily. From time to time, goats thought to be unregistered are found to be earmarked, and they can be registered when the earmark is checked.

Natural rearing on the dam

The natural way to rear a kid is to allow it free access to suckle its mother until either the mother stops the kid feeding, or she goes dry. If the goat-keeper's time is limited, this can be a successful way of rearing kids. However, there are several drawbacks. Firstly, the goat-keeper has no idea how much milk the dam is giving or the kid is taking. A greedy kid may well overfeed itself at the expense of bulk food, or a weak kid may not get sufficient. If a goat is feeding two or three kids, she may not have sufficient milk to feed them adequately and may tend to favour one kid at the expense of the others. Since the kid requires less milk as it gets older, it will not empty the goat's udder so frequently and this will cause the dam to start drying off. This can be overcome by the goat-keeper ensuring that the goat's udder is completely stripped out at least once, preferably twice a day. The goat may decide that the milk is intended only for her kids and may refuse to let down when milked. A successful modification of natural rearing is, when the kid is a week old, to put it in a separate pen at night, milk the goat in the morning then let the kid run with the dam in the day.

If natural rearing is attempted, it is worth setting some time aside each day to handle the kids, otherwise they may well grow up shy of people and difficult to manage. Bottle-fed kids are handled several times a day, so tend not to be apprehensive of people. Where the intention is to produce optimum milk yields from the dam and sturdy kids for herd replacement or sale, natural rearing is not recommended.

Rearing on bottle or bucket

Even if kids are to be artificially reared, it is normal practice to leave the kids with the dam for approximately the first four days while the goat is still giving colostrum. Many goat-keepers give kids that are to be bottle fed their first feed from a bottle within the first half-hour after birth. This not only gives them the idea of bottle feeding right from the start, but also ensures that they have colostrum. It is absolutely essential to the kid's survival that it has plenty of colostrum in the first six hours after birth.

At about four days of age, the kids are separated from the dam by being put in a separate pen. It is best if this is done while the dam is out grazing, rather than in her presence. It is advisable to keep the kids away from their mother for at least two weeks for successful weaning. After separating them from their dam, the kids should be left for three or four hours to get hungry, then take a bottle of warmed milk (blood heat), preferably their own mother's milk, and offer it to the kids. The bottles need to hold 1¼ to 1⅓ pt (0.75 litre) (wine or squash bottles are ideal), and a teat suitable for lamb feeding put on. (The holes in teats are nearly always too small and need to be enlarged).

To start the kids drinking, the kid's head is held from above by putting the palm of the hand over the kid's face and slipping fingers and thumb into the kids mouth to open it slightly. The teat is slipped into the kid's mouth, and the kid encouraged to suck. Most kids get the idea within two or three feeds, but more perseverance may be needed.

Buckets are usually too big to start a kid on but milk can be put into a bowl and the kid encouraged to drink by putting a milky finger in the kid's mouth and gradually drawing (never pushing), its head towards the milk. Again, it may take two or three feeds.

The kids may take about 0.12 litres (¼ pt) at first, and should be given four feeds a day until their appetite has increased to take approximately 2 litres (4 pt) a day. From then on, three feeds a day are sufficient. A male kid will benefit from receiving 5 pt (2.5 litre) a day, for the extra growth he will need to make. Current practice in the UK is to bottle feed until the kids are between four and six months old. Weaning is carried out gradually by cutting down, then altogether removing one feed, then another and finally ceasing milk feeding altogether. As the kid gradually receives less milk, it will compensate for the loss by eating more bulk foods and concentrates.

Milk substitutes

The best milk to feed a kid is its mother's milk since this contains antibodies to protect the kid against disease. In addition, goats milk is a perfectly balanced food. However, two kids will drink about 4.5 litre (1 gal) a day, which probably leaves very little spare from the average household milker. Milk substitutes designed for kids, lambs or calves can be used, so long as the kid has been started on its mother's colostrum and preferably has been fed upon goats milk for a week or two. Dried milks prepared for lambs seem to give the best results.

The instructions for lamb milks usually recommend starting lambs on half strength milk, then doubling the strength after a few weeks. Since goats milk has about half the butterfat content of ewes milk, kids do not require the doubling in strength and will grow very satisfactorily on the more dilute milk. The milk must be mixed according to instructions and care should be taken to ensure that it is free from lumps, which might cause digestive upsets.

Early weaning

Early weaning has become fashionable with calves because of economic pressures. If a good market exists for all the milk, a goat-keeper could seriously consider early weaning. The kids are milk fed by one of the above methods for the first two to three weeks, and offered calf weaning pellets *ad lib*. The milk feeding is rapidly cut down and the kids depend on the high protein concentrate feed and lots of high quality hay. This is a very successful method but care should be taken on the economics as the calf pellets are expensive. Obviously this method saves time. Kids should, at all times, be offered soft clean hay and bulk green foods. Concentrates can be introduced after the first week (a little flaked maize can be useful to tempt the kids).

Weaning

A naturally-reared kid may well begin to lose interest in suckling after four or five months of age, and may wean itself. However, many try to carry on feeding. The best way to wean is to shut the kid away from the doe, firstly at night only, then all day as well, for at least a fortnight until it no longer wants to feed from its mother. This process is often more traumatic than taking the kids off at four days, since a four or five month old makes a more vociferous and prolonged objection.

Weaning from a bottle or bucket has already been described, but it must be remembered that during weaning the milk feed must be replaced by high quality bulk foods to prevent a serious arrest in growth.

It is often the practice to keep artificially-reared kids inside until they are weaned. They are put out to graze with the herd when they are approximately six months old. Kids should be put onto clean pasture since they are particularly susceptible to worm infestation, and should be checked carefully for signs of scouring and loss of condition. The kids should be wormed if there is any suspicion that an infestation is building up.

Kids should be encouraged to eat large amounts and varieties of bulk foods from as early as one week of age to develop the appetite and digestive capacity that will be so important when they become milkers.

9 Milk and milking

Composition and characteristics

Goats milk, which we may define as the normal secretion of the healthy udder, is similar in general composition to cows milk. Both consist of about 90% water with other materials (of which butterfat and protein are major constituents) dissolved and suspended in it. There is no great difference in flavour between the two milks, assuming of course that both have been produced and stored with attention to hygiene and that the goats whose milk is sampled are not suffering from a deficiency of dietary cobalt.

One sometimes hears of people who mistakenly consider that goats milk is 'richer' than cows. This is not so. The level of butterfat (responsible for the creaminess of the milk) is of the same order and shows the same sort of range in both cows and goats (between about 3% and 6% by weight). In both species the milk composition, and particularly the amount of butterfat present, varies due to the stage of lactation of the animal (fat levels are high immediately after kidding, fall as the lactation becomes established and rise again as milk yield declines), the time of the year (fat levels tend to be low in spring and early summer when the better weather results in lush pasture growth) and the animal's diet. Hereditary factors and the breed of the animal also affect composition and butterfat content. Thus, on average, an Anglo-Nubian will tend to have milk with a higher butterfat level than will a Toggenburg, but within each breed there will be families having higher or lower levels than the average. Of all factors connected with milk yield and composition, butterfat content is the one most easily altered by a planned breeding programme. The combination of economic pressures and the widespread availability of semen from premium bulls has resulted, in the last decade or two, in a general increase in average butterfat levels and a reduction in the range of butterfat values in cows milk. These factors have not been felt to anything like the same extent in the goat world, with the result that the range in butterfat values between goats is much wider than between cows.

Goats' milk differs from cows' in several distinct aspects. Firstly, the fat particles are much smaller and more evenly distributed through the milk. As a result, cream rises more slowly to the top than it does in cows milk. Secondly, goats' milk and milk products are pure white in colour. If yellow butter and cheese are required, the colour has to be added in the form of annatto. Thirdly, the curd formed by coagulation of the milk is very much softer than that from cows' milk. Fourthly, the salt level is approximately two and a half times that of cows' milk. Finally, the milk is more highly buffered, with naturally occurring chemicals.

Medical aspects

Goats milk and products play an important role in the treatment of certain allergic conditions, such as infantile eczema and asthma, where these are due to a reaction to cows' milk and bovine protein. However, it is important to remember that goats milk is not the cure for such conditions, the cure is to stop consuming the allergenic material. All that goats milk can do is to provide nutritionally suitable substitutes for the cows' milk and nutrients missed out from the diet.

The softer curd and greater natural buffering of goats' milk make it more easily digested than cows' milk with the result that it is ideal for infant feeding and for persons with disorders of the gastro-intestinal tract, such as gastric and duodenal ulcers. It was sometimes said that infants fed on goats milk developed anaemia as a result. This is not wholly true. Any baby fed on other than human milk (e.g., cows) for prolonged periods will develop anaemia since all such natural milks are deficient in iron. Fed as part of a sensible diet, goats milk will not cause any problems.

Goats' milk, in Great Britain at least, has rarely been associated with the spread of disease. In some parts of the world it carries the bacterium *Brucella melitensis* responsible for the disease known as Malta or undulant fever in man. This disease is not known in goats in Britain. Goats are not usually subject to the *Brucella* organism of cattle (*B. abortus*), which causes contagious abortion in cows and also undulant fever (brucellosis) in man. As a result, goats are not affected by legislation under the current British brucellosis eradication scheme, although they are included in schemes in some areas of the United States.

Whilst it is not possible to say that goats in Britain never carry tuberculosis, the numbers likely to be affected are small and very much less than the numbers of cows that carried the disease before the Tuberculin attestation schemes earlier this century.

For these reasons, it is not essential to pasteurize goats milk for domestic consumption. However pasteurization will improve keeping qualities and may be advisable where the milk is for sale.

Milking

The art of hand milking is usually learnt quickly and thereafter becomes an almost subconscious act. Sit at the goat's side, either facing its flank or looking towards the tail, in whichever position is most comfortable. (It may prove easier if the goat jumps up onto a milking stand). Grasp a teat between the thumb and first finger of each hand. Close the finger and thumb of one hand together to stop the milk being forced back into the udder and squeeze the milk from the orifice of the teat by closing the fingers progressively around the teat from the top. When all the milk has been squeezed out relax the grip of thumb and first finger (without letting go of the teat) to allow the teat to refill. Repeat the process with the other hand and teat and proceed alternately in this manner until the milk flow diminishes. Remove the hands from the teats and massage the udder, from the top downwards, for a few seconds and then milk as previously described. Repeat the massaging and milking until the flow ceases and all the milk has been removed.

The essential feature of hand-milking goats is that the milk is removed by squeezing the teats. There is no need to pull the teats, or to

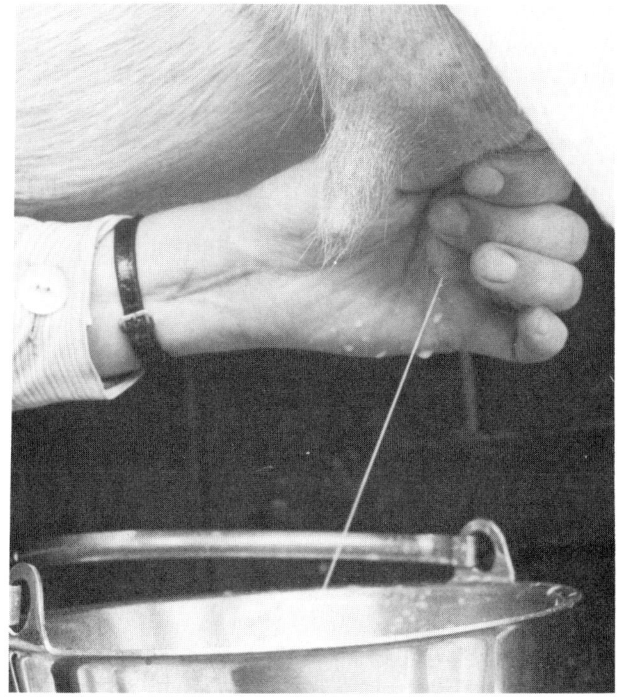

Hand milking, showing
the action of the fingers.

run finger and thumb down them even to strip the final volume of milk from the udder. To do so may encourage the teats to become enlarged and elongated and hence increase the risk of them being damaged.

Let-down

Let-down is the bodily mechanism whereby the flow of milk is stimulated at each milking time. It is caused by the release of a hormone into the blood stream in response to nervous stimuli. Events that may stimulate let-down include, for example, the rattle of the milking pail, the sight of the previous goat to be milked being returned to her pen, udder washing and so on. The effect of the hormone is short lived. If milking is not completed within about five minutes or so of stimulation of let-down, any milk remaining in the udder is unlikely to be removed.

For optimum milk yields, the goat-keeper must develop and maintain a routine at milking time. Changes in routine, such as alteration to the order in which goats are milked or a different person milking the goat, can result in let-down before milking or its total suppression. In either case, yields will suffer.

Milking interval

It has been shown that optimum milk yields are produced when the intervals between successive milkings are kept constant and as near equal as possible. The ideal arrangement would be for twice daily milking at equal intervals of twelve hours, but other activities (such as looking after the family and going out to work) tend to mean that unequal intervals are likely to be more convenient. Alternating intervals of fourteen and ten hours should be regarded as the safest maximum variation from the ideal. Constancy of milking intervals becomes of less importance later in the lactation when yields have declined.

Equipment

Equipment for milking, and straining and cooling milk must be purchased or improvized from items found in the kitchen. In selecting equipment, the nature of the material from which it is made is of prime importance. Stainless steel is by far the best. Although initially costly, it is durable and, perhaps more importantly, has a surface that has less microscopic depressions in which milk residues and the microorganisms that cause souring and off-flavours to develop, can build up. Tinned dairy equipment is satisfactory, except for the fact that the tinning resulting from present methods of manufacture does not stand up to repeated use of chemical sterilizers. Plastic items are easily scratched and allow milk residues to accumulate. Only food grade plastics should be used, for reasons explained in Chapter 10.

A strip cup can be purchased to collect the foremilk in, although any suitable receptacle (such as a discarded cup) can be pressed into service since the foremilk will be thrown away. A cup of a dark colour will make examination of the foremilk easier.

Although milking machines suitable for goats are now more readily available, they are costly, require considerable time to dismantle and clean and are, therefore, unlikely to be used unless more than a dozen goats are to be milked. For smaller numbers, hand milking using a stainless steel bucket of about 7 litre (1½ gal) capacity is the alternative.

Newly drawn milk should be strained immediately. A purpose-made milk strainer can be purchased, which will fit the neck of a churn. These are funnel-shaped and contain a pair of perforated plates between which a milk filter can be sandwiched. A strainer can be improvized from a pair of nylon sieves and a disposable wet strength tissue can be used instead of a milk filter.

A cluster and bucket for machine milking, and a strip cup, photographed in the dairy. Note the dairy wash tanks in the background.

A portable milking machine.

Churns or hand-cans of various sizes are available for storing milk. The larger ones are usually made of tinned or stainless steel and hand-cans of aluminium. Where only a few goats are kept a number of 4.5 litre (1 gal) or 2.25 litre (½ gal) hand-cans would be ideal since they will fit into the average domestic refrigerator.

Once in the churn, milk must be cooled quickly. The only satisfactory method involves the use of cold, running water. Hand-cans or smaller churns can be stood in a bowl of water with cold water from the tap trickling into it. An alternative is to make a perforated hose into a ring, leaving a connection to be fitted with another hose from the tap. The ring is slipped over the neck of the churn and water allowed to trickle down the sides. For larger churns, an in-churn cooler will be needed. This is, essentially, a paddle through which cold water circulates and which is driven round by water pressure.

A balance, to weigh the milk given by each goat, is essential. Balances designed specifically for the purpose are available. Alternatively, the kitchen scales can be used although it does make life simpler if the scales or balance can be permanently adjusted to allow for the tare of the empty bucket.

Quality milk production—cleanliness and hygiene

The production of goats milk requires considerable time and effort. With little more effort and attention to detail, the producer can ensure that the milk reaches the consumer, whether customer or family, in the best possible condition.

Personal hygiene is of prime importance. The person doing the milking must wear a clean overall or other suitable protective clothing. Hands and forearms should be washed before milking and nails scrubbed and regularly trimmed. Milking should not be undertaken by anyone suffering from a contagious medical condition such as sickness, diarrhoea, or boils. Cuts and grazes should be covered with a waterproof dressing. Smoking, of course, should not be permitted whilst goats are being milked or the milk is being handled.

Milking should not take place in the goat's pen. A separate parlour, used for no other purpose, would be ideal. Where this is not practicable, goats should be milked in a part of the goat shed that can be

An in-churn cooler, which consists of a paddle through which cold water circulates. It is essential for cooling milk in larger churns.

Milking equipment. L to R: 4·5 litre (1 gal) and 2·25 litre (½ gal) hand cans, strainer and stainless steel milk bucket.

washed down before and after use. Wherever goats are milked, steps should be taken to ensure that the atmosphere remains as free from dust as is possible. Principal sources of dust are food and bedding materials. No hay or straw should be moved around in the shed prior to milking and concentrates must not be fed before or during milking.

When the goat is brought to the milking area, debris that might fall into the milking bucket should be removed from her flanks and udder. To avoid creating dust this is best done with a very slightly moistened cloth (which is not that used for udder washing where this is done routinely). Loose hair can be removed by regular grooming with a brush, but this should not be done just prior to milking.

When cows are milked, it is normal practice for the herdsman to wash their udders with a solution having detergent and bactericidal properties. Opinions are divided as to the necessity for this with goats, principally because of the obvious difference in consistency of goat and cow dung. Udder washing will be essential where a goat has been scouring or has walked through muddy field gateways. When udders are washed, a clean cloth should be used for each goat: disposable cloths discarded after use prevent the build-up of bacteria. The udder should be dried before milking commences. Washing will remove protective oils from the skin of the udder. A good-quality udder cream containing a bactericide should be applied after milking, for reasons of udder health rather than milk quality.

The first two or three squirts of milk from each teat should be directed into a strip cup and not into the milking bucket, for two reasons. First, it allows the foremilk to be examined for the presence of clots, blood or other abnormalities which could be indicative of mastitis. Second, the foremilk has a much higher bacterial load than the remainder of the milk, and would impair the keeping quality were it included.

Straining the milk through a disposable filter will remove any sizeable debris (such as hairs and hayseeds). It will not remove bacteria. Newly-strained milk must be cooled as rapidly as possible to a temperature at, or below, 4°C (40°F) and then stored at this temperature. This will prevent the multiplication of bacteria, which otherwise will cause off-flavours to develop and reduce keeping quality. Cooling must be carried out using cold, running water and then, when the milk has cooled to the temperature of the water it is placed immediately in a refrigerator. Simply to stand a container of warm milk in a refrigerator will not lower its temperature rapidly enough, although a refrigerator is the ideal place in which to store milk that has already been cooled.

Taints in milk

Milk absorbs odours easily and should not be stood or stored near to strong smelling things such as oranges, creosote, disinfectants and so on.

Taints in milk can also result from the consumption of certain foodstuffs (silage, fish meal, excessive amounts of cabbage or kale, etc.), from mineral imbalance (particularly cobalt deficiency), from disease (mastitis can result in a salty taste to the milk) and from inadequate cleaning and sterilizing of equipment.

A major cause of taints is the action of naturally occurring enzymes in the milk, which can release the fatty acids, capric, caprylic and caproic, which result in a 'goaty' flavour. Heat treating the milk to 57°C (135°F) will denature the enzymes and prevent this.

Taints may also be of genetic origin, and these cannot be eliminated.

Cleansing and sterilization of equipment

All equipment used for dairy purposes should be rinsed thoroughly in clean water immediately after use. The temperature of the water is critical. It should be tepid, that is about 46°C (115°F). At temperatures over 52°C (125°F) milk protein is denatured and deposited on surfaces of the equipment in a film that is difficult to remove. Cold water will cause the butterfat to harden into a greasy deposit, rendering chemical sterilization ineffective.

After rinsing, equipment should be washed thoroughly using warm water and detergent. A brush with nylon bristles can be used to scrub the utensils but scouring pads or powders, which might cause scratches, should never be used. All traces of detergent should be removed by thorough rinsing in hot water prior to sterilization.

Although heat, in the form of steam, is used sometimes to sterilize equipment in large commercial dairies, chemical sterilization is now the commonest method and the most practicable for domestic use. Of the approved sterilizers, the most readily obtainable is sodium hypochlorite, which is available under various trade names from agricultural suppliers. (It is also available in a combined form with detergent). It must be freshly prepared each time it is to be used. It should be diluted according to the instructions on the pack and the equipment filled with the solution, or, wherever possible, immersed in it for ten minutes. After sterilizing, the equipment should be rinsed in clean water and allowed to dry by drainage, upside down in a dust-free place. Any items, such as brushes and cloths used to wash equipment, should be sterilized also.

If, in spite of the twice-daily washing and sterilizing routine described above, a build-up of the milk deposits is seen, the equipment should be treated with a proprietary milkstone remover.

The milk from sick animals and those receiving veterinary treatment should be discarded, although it may be possible in the latter case for it to be used for stock feeding. Check with the veterinary surgeon or the drug manufacturer's instructions. Always milk animals receiving drugs or with contagious illnesses and particularly conditions affecting the udder last of all and wash your hands well afterwards.

Packaging

When milk is sold, even in small amounts to neighbours, it should be packaged correctly in disposable containers. Two forms of packaging are available: waxed cartons sealed by heat or metal clips, and polythene bags sealed by heat. Bags and cartons, preprinted with the nature of the contents in attractive designs are available. Re-used glass or plastic bottles and jars are not suitable.

Freezing

Goats milk, unlike cows milk, freezes successfully, thus allowing constant year-round usage in spite of seasonal variations in supply.

Milk should be placed in containers as soon as possible after straining and put immediately into the fast-freezing compartment of the freezer, to reduce the likelihood of the cream rising. Once frozen, it may be stored elsewhere in the freezer but should never be allowed to become warmer than the usual temperature that the freezer runs at. It is most successful if milk is frozen in fairly small quantities, for example, 1 pt cartons or bags. To use, it should be allowed to defrost naturally either at room temperature or, preferably, in a refrigerator.

Selling milk and dairy products

The production of cows milk is governed by the Food and Drugs Act 1955, and the Milk and Dairies Regulations 1959, which are administered by the Ministry of Agriculture. For reasons that no doubt seem logical to the bureaucrat, goats milk, and produce derived from it, is not subject to the Milk and Dairies Regulations. In other words it is treated as a food rather than as milk.

In practice, this means that the goats milk producer is subject to fewer controls than the cow dairyman, but since the Food and Drugs Act is administered by local authorities, interpretations of the Act can vary from place to place. The principal provision of the Act is that it is an offence to offer for sale any commodity which is not of the nature, substance or quality demanded or which has been adulterated, diluted or is unfit for human consumption. As a result, containers must be labelled with the nature (goat's milk, not just milk) and amount of their contents. The producer's name and address must also appear on the container where the product is to be sold to the public via an intermediary, such as a shop. (Packaging and labelling are the responsibility of County Council Consumer Protection Departments). Local Authorities differ in their insistence upon separate facilities for handling milk and the zeal with which they inspect such facilities. However, there are strong indications that the authorities are taking goats milk production and selling much more seriously than they did in the past. In view of this, advice should be sought from the Environmental Health Department of the local District Council before any sales are made, even if these are to be on a restricted or casual basis. This Department is also able to arrange microbiological testing of milk on request.

Recording yields

A record of the daily milk yields of all goats in the herd is an essential management tool. It will allow the owner to cull unproductive stock and keep, and breed replacements from, the better producers. It provides evidence of the effects of changes in management of the herd, such as alterations in feeding practice, and thus helps the owner to develop the most suitable system for his own stock. The milk yielded by an individual goat is a sensitive indicator of her bodily state. A sudden drop in yield can be the first sign of incipient illness or the onset of heat.

It is usual practice to weigh milk because of the difficulties in attempting accurately to measure the volume of freshly drawn frothy milk.

Although a private record of yields is perfectly adequate for purposes of herd management, an official record made by an independent body is essential for the serious pedigree breeder. Members of the British Goat Society can arrange, through the Society, for their goats' yields to be recorded by the Milk Marketing Board in the same manner as dairy cattle. The MMB service is relatively expensive. A cheaper alternative available in certain parts of Britain is Club Milk Recording. It is organized by local goat-keeping societies affiliated to the BGS and subject to the approval of the parent society. Goats recorded under either the MMB or Club schemes can qualify for awards in the form of prefixes indicating their yields in a lactation period of 365 days, thus providing official recognition of their productive capabilities. (Full details of the appropriate conditions can be found in the British Goat Society's Regulations). Such official recognition is of benefit both to the owner of the goat, who might wish to sell or export progeny, and to goat-keepers in general, who need to establish the whereabouts of productive lines if they are to breed for an overall improvement in milk yields. It would be of great benefit to goat-keeping in Britain if some of the money and effort expended upon showing was diverted to official milk recording.

10 Dairy and other products

Goats milk can be used for making a variety of dairy products, and little modification to recipes designed for cows milk is necessary. Much of the equipment required can be improvized using ordinary domestic items. A word of warning is necessary here. Any materials used for dairy equipment should be of food grade, and suitable for coming into contact with food. Plastic tubs that are used for jams and marmalade make excellent cheese moulds when their bases have been perforated, but similar tubs which have contained putty, glue, and so on, are not suitable. Impurities can leach from the plastics into the product and lead to a slow accumulation of toxic substances in the consumer. Similarly, moulds for hard cheeses can be made from plastic water piping, but never from drain piping.

Some recipes for various dairy products are given below. Whilst these are by no means exhaustive, they give some idea of the range of delicious produce which can be made. They are intended for the goat-keeper who wants to process milk for domestic purposes, and are in no way meant to be for large scale commercial use.

Care of equipment

Dairy utensils must be kept clean, and must be sterilized immediately before use. All plastic, glass, stainless steel and enamelled equipment and metal spoons can be sterilized in hypochlorite solution (see Chapter 9). Any tinned utensils should be scalded with boiling water. Cheese cloths are rinsed, washed in soapy water then boiled after use, then boiled again for ten minutes immediately before use. These procedures help to prevent the growth of bacteria that may adversely affect the dairy products.

Cream

Despite the fact that the cream rises to the top of the milk more slowly than that of cows, the simplest way to take the cream from goats' milk

Cream and butter making equipment. L to R: Cream separator, fleet, butter churn and Scotch hands.

is to hand skim. Fresh milk is set in a shallow bowl or setting pan, covered with a cloth, and left in a cool place for approximately twelve hours. The cream is skimmed off with a fleet, a metal saucer-like implement with small holes in the centre to allow the milk to run away. Special setting pans are also available with a lip to allow the milk to be poured off from under the cream.

Hand skimming methods produce single cream, and are not as efficient at removing all the cream as mechanical methods. However, the cream produced is well suited to domestic culinary use and is of an ideal thickness for butter and ice cream making. Hand skimming is probably most successful with milk of high butterfat content, such as that from Anglo-Nubians.

A more efficient method of removing cream, and the only method for obtaining double cream, is to use a separator. Hand or electric models are available either new, or occasionally second hand. Milk at blood heat is poured into the bowl, from which it runs down into the machine and is forced through a series of perforated conical plates by centrifugal force, created either manually (by turning the handle) or by an electric motor. The cream is forced out through one spout, and the skim milk through another. The thickness of the cream can be altered by adjusting the cream screw on the side of the cone. It is not worth setting up a separator unless a minimum of 10 to 15 litre (2 to 3 gal) of milk is to be separated, as the time spent is barely justified for less.

Butter

Butter can be made with milk soured by a starter culture. However, for home consumption, 'sweet' cream (fresh, unsoured cream) makes a very acceptable product. Butter can be made using a glass butter churn or an electric mixer. The only other equipment required is some muslin, a thermometer (a dairy thermometer is ideal although a sugar thermometer is suitable), a wooden board and a pair of Scotch hands. The board and Scotch hands should be scalded before commencing, and the Scotch hands stood in a jug of brine.

Hand-skimmed cream is ideal, or alternatively, mechanically-separated cream that is thinned with tepid water until it runs freely over the back of a wooden spoon. At least 0.5 litre (about 1 pt) and preferably more should be used, since the yield will otherwise not be worthwhile. The cream needs to be brought to the correct temperature, which is between 14°C (56°F) and 17°C (62°F). Use the higher temperature if the room is cold and the lower if warm.

The cream is put into the jar or mixer bowl and churned, or mixed with a cake mixer attachment, on a slow speed, until grains of butter begin to appear, and the liquid begins to look watery. This liquid is buttermilk, and is drained off, and water, a little colder than the churning temperature is added, sufficient to float the grains. The paddles of the churn are rotated slowly a few times, or, if using a mixer, the grains are stirred using a spatula. The water is strained off and the grains tipped into a colander lined with muslin and put under cold running water. The grains are moved around gently in the muslin until the water runs clear, and then left to drain. They are turned onto a board and sprinkled with one teaspoon of salt for every 450 g (1 lb) of butter. The grains are then pressed together with the Scotch hands and worked and squeezed to remove the water. When as much water as possible has been removed, the butter is patted up into rectangular shapes or rounds, wrapped in greaseproof paper and stored in the refrigerator. This butter will come out pure white. If a yellow colour is required, butter annatto is added to the cream before churning.

Ice cream

Cream does not freeze successfully without some form of processing and various soufflés, fools and mousses can be made as a means of preserving it. Ice cream is an ideal way to store cream.

Half a litre (1 pt) of single cream is required. A syrup is made by dissolving 75 g (3 oz) of sugar in half a cup of water, bringing to the boil, and boiling for five minutes. Meanwhile, three eggs are separated and the yolks beaten. The boiling syrup is poured onto the yolks and the mixture whisked until pale and frothy. The cream, to which 0.5 teaspoon of vanilla essence has been added is stirred in, and the mixture poured into a plastic tub or equivalent and placed in the freezer. When frozen to a slush, the egg whites are whisked until firm and folded in and the ice cream frozen until solid. The tub should be moved to the refrigerator approximately one hour before the ice cream is to be served.

Yoghurt

Yoghurt is milk soured by a bacterial culture. It can be made simply, with the minimum of equipment. It has a wide variety of uses, not only as a dessert with fruit, but also in savoury dishes, in salad dressings and as a replacement for cream in many recipes.

The equipment required is a wide-necked thermos flask or electric yoghurt maker, a thermometer (dairy or sugar), a pan and a pot of unsweetened natural yoghurt to act as starter culture. All equipment is sterilized with hypochlorite solution before use. The required amount of milk is heated to 82°C (180°F) then cooled by placing the pan in a bowl of cold water. At 49°C (120°F), the culture is added (one tablespoonful per pint of milk). After thoroughly stirring, the pan is reheated until the milk is again at 49°C (120°F). It is poured into a preheated flask or the cartons, and the flask set in a warm place, or the yoghurt maker closed and switched on. After approximately eight hours, the yoghurt should be made and is placed in the refrigerator to cool. This yoghurt can then be used to start a further batch, but the culture should be replaced from time to time to maintain the correct bacteria and to avoid contamination by yeasts.

Thick yoghurt comes from milk with high total solids, consequently Anglo-Nubian milk tends to make better yoghurt than that from the Swiss breeds. However, dried milk can be stirred into the yoghurt if necessary to thicken the curd.

Cheese

Goats milk does not sour naturally with the same ease as cows milk. Consequently it is difficult to produce a good goat cheese without using a lactic starter. Lactic starters can be obtained from dairy suppliers. Using the starter culture, most of the recipes for cows milk cheeses can be used with very little modification.

Yoghurt cheese

This is made very simply by turning yoghurt into a boiled cloth (cheese cloth, double thickness muslin or cotton sheeting are ideal) and hanging to drain. After twenty-four hours, the curd is salted with one level teaspoonful of salt per 450 g (1 lb) of curd, and placed in a fresh cloth. After a further twenty-four hours, the cloth is removed and the cheese stored in a bowl in the refrigerator. This is a mildly acid soft cheese.

Crowdie

This is a soft cheese with a sharp acid flavour. 4.5 litre (1 gal) of milk is heated to 71°C (160°F) and cooled to 27°C (80°F). Two tablespoons of starter are added and stirred in thoroughly. Two to three drops of cheese rennet (not junket rennet) are diluted in an eggcup of water and also stirred in. After approximately one hour, a firm curd should have formed. The container of milk is placed in a bowl of hot water to raise the temperature of the curd to 38°C (100°F). Whilst warming, the curd is gently sliced with a perforated spoon. When 38°C is reached, the curd is kept at that temperature for thirty minutes.

The curds are poured into a boiled cloth, which is hung up to drain, and the salting and changing of cloths carried out as for yoghurt cheese. The curd should be firm enough to make into rounds or rolls.

Both yoghurt cheese and crowdie freeze well, and so are useful ways of storing a summer surplus of milk.

Hard cheese

Hard cheese requires some specialized equipment in the form of curd knives, cheese moulds and a press. Its manufacture is probably best learnt by watching a proficient cheesemaker. For these reasons, recipes are not given here. Detailed instructions and recipes are to be found in several excellent cheesemaking books (see Bibliography).

Meat

Goat meat in Britain is usually referred to as 'kid', but in the USA the term 'chevon' is used. Usually only male kids are reared for meat, since most females are used for breeding. However, surplus females and those with congenital faults are suitable for meat production. Meat kids can be reared on goats milk or on milk substitute (lamb or calf milk). The substitute is usually the most economical method. Concentrate ration is introduced as for breeding stock, and the kids can be weaned early, at three to four weeks old (Chapter 8, Early Weaning).

As long as male kids are to be slaughtered before six months of age and can be kept separate from female kids, castration is not necessary.

However, if the male kids are run with breeding females they must be castrated if the intention is to keep them beyond twelve weeks of age. Castration can be carried out using a rubber ring applied with an 'Elastrator' device, or surgically by a veterinary surgeon. Castration should be carried out early in the goat's life, and it is better to apply a rubber ring at the same time as disbudding, but never later than one week of age.

Kids can be slaughtered at any age. Within the first week after birth, the carcass is small, usually cooked whole, and somewhat like chicken meat in taste and texture. By twelve to sixteen weeks the carcass is similar to a small lamb and by six months should be the size and texture of a large lamb or small deer. It is usual to have kids butchered to give the same joints and cuts as lamb. The meat is extremely versatile and, because it is not interlarded with fat, is lean when trimmed.

Slaughtering may be carried out at home, if done humanely, on condition that the meat is not for sale and is only for home consumption. The most humane way is to use a captive bolt pistol (for which a firearms certificate is required) to stun the animal before cutting its throat. Otherwise, the animals can be slaughtered at abattoirs. Some abattoirs arrange butchering as well, and others return the carcass for the owner to dispose of. If the meat is intended for sale, it must be inspected at the time of slaughter.

Skins and curing

If skins are required for curing, the slaughterman should be told beforehand to ensure that he removes the skin with the minimum of damage. Goats are extremely difficult to skin and must be done whilst the carcass is still warm.

The best results are obtained if skins are tanned professionally, but there are several successful methods which the goat keeper can use himself.

Method 1

400 g (14 oz) alum and 150 g (5 oz) salt. Dissolve in 5 litre (9 pt) hot water; cool to 37°C (99°F)

Scrape all the fat off the skin using a blunt knife and immerse the skin in a bucket of the solution. Soak for 48 hours, agitating the skin from time to time. Remove the skin from the solution and dry slowly in the fresh air, 'working' the skin frequently to prevent it going hard. When dry, trim round the skin from the underneath using a sharp razor blade and gently brush the hair until smooth.

Method 2

Cooking salt and 4 litre (1 gal) paraffin (kerosene)

Cover the inside of the skin with salt as soon as possible after removal from the goat and fold the skin over on itself to prevent drying out. After one to two days, wash the skin in clean water, and nail it out on a board, hair side down. Scrape with a blunt knife to remove fat, then sprinkle paraffin all over the skin and rub in with a cloth. Apply paraffin twice more (at three day intervals). As the skin begins to dry, rub with sandpaper. After the third application of paraffin, remove the skin from the board and work it thoroughly and frequently to make it supple. Trim as in method 1.

Method 3

Excellent results are obtained using 'Lankroline', a tanning oil (see Further Information: Suppliers). Instructions are included.

Manure

Well-rotted goat manure is excellent for the garden, whether dug in or used as a mulch. It must be well rotted to prevent hay seeds in the manure from germinating. As soiled litter from goat pens is very dry compared with that of cows or pigs, it is often advantageous to mix some wetter muck in with it to ensure faster and more complete rotting. In a dry spell, wetting the heap will help to speed up decomposition. Sales of bagged goat manure can give a small but steady income to the goat enterprise.

11 Health care

Preventative measures and techniques

Hoof trimming

In the wild, the rough stony places which are the goat's natural habitat
keep their hooves worn down. Goats spending a lot of time on a concrete exercise yard will similarly wear down their hooves and not
require as much attention as a goat spending its time penned on soft
bedding or out at pasture. The goat is cloven hoofed (two 'claws' on
each foot) and each claw consists of a soft sole with a horny outer rim.
Both parts of the claw grow, but the horny layer grows faster than the
sole. Hooves require trimming every one or two months, depending on
the ground on which the goats exercise. Foot rot shears or a sharp knife
(e.g., Stanley with interchangeable blades) are suitable.

The goat is tied firmly against a wall. The attendant stands alongside the goat facing towards its tail. The foreleg is lifted and secured
firmly between the herdsman's own knees. The outer horny part of the
hoof is pared away until it lies flat with the soft sole. The heel part of
the sole is then sliced off so the whole walking surface of the hoof is flat.
A rasp or file (e.g., Surform) can be used to smooth off the hoof. All
hoofs are trimmed similarly, and the goat walked for a short distance to
check that the hoofs are level and that it moves soundly.

Overgrown hooves not only result in the goat walking badly, they
also encourage foot rot, which can spread rapidly through a herd causing severe lameness.

Vaccinating

Goats, like sheep, are subject to several clostridial diseases, in particular tetanus and enterotoxaemia. The diseases are nearly always fatal.
Protection by vaccination is highly effective and relatively inexpensive.
Vaccines can be obtained from agricultural merchants and can be
administered either by the goat owner or by a veterinary surgeon.
Administration is simple, by subcutaneous injection, but it would be
unwise to attempt it without first receiving some instruction. Absolute
cleanliness is essential. A sterile syringe and needle must be used.

An in-kid doe is vaccinated two or three weeks before kidding to protect herself and her kids. Antibodies in colostrum and milk will pass immunity on to the kids after birth. Kids should be vaccinated at two to three months of age, followed by a further injection approximately one month later. Thereafter, single annual vaccinations are required.

Worming

Routine treatment against roundworms is advisable for all goats. Goats, being browsers by nature, do not have the same natural resistance as that of sheep to roundworms, which spend part of their life cycle on pasture grass. Worm infestation causes loss of condition, stunts growth of young stock and impairs milk yields. There are many worming formulations on the market. Those used for sheep or cattle are suitable, but it is essential to ensure that they can be administered to milking stock. The milk can be fed back to kids, but must not be used for human consumption for the stated length of time.

Some worming compounds are injected, some are administered by oral drench, and some are tablet or in-feed pellets. Availability in small quantities is often the factor that determines which type to choose, although some goats will reject in-feed formulations. Since the various worming preparations on the market are active against different ranges of worms, and since there is some evidence that worms might become resistant to certain preparations, it is prudent to vary the drug used.

As a routine, worming should be carried out during the first week after kidding, before first turning stock out to grass in the spring, and again in the autumn at the end of the grazing season. (Stall fed goats should be wormed at least twice a year). It is also advisable to worm in the early summr after a warm wet period when worm larvae build up for a massive re-infestation of the goats. At any time when a goat looks out of condition for no apparent reason, the droppings appear clumped together rather than as separate pellets or the milk yield suddenly drops, it is worth worming to be on the safe side.

Delousing

Particularly during the winter, goats are susceptible to lice, which are tiny arthropods living as parasites on the skin. They cause irritation and can result in general debilitation. Regular, thorough dusting with a proprietary louse powder (usually containing BHC) every two to three months will keep the goats' coats clean and free from parasites.

Drenching

Drenching is the method of administering liquid medicines by mouth to animals. It can be done either with a narrow-necked drenching bot-

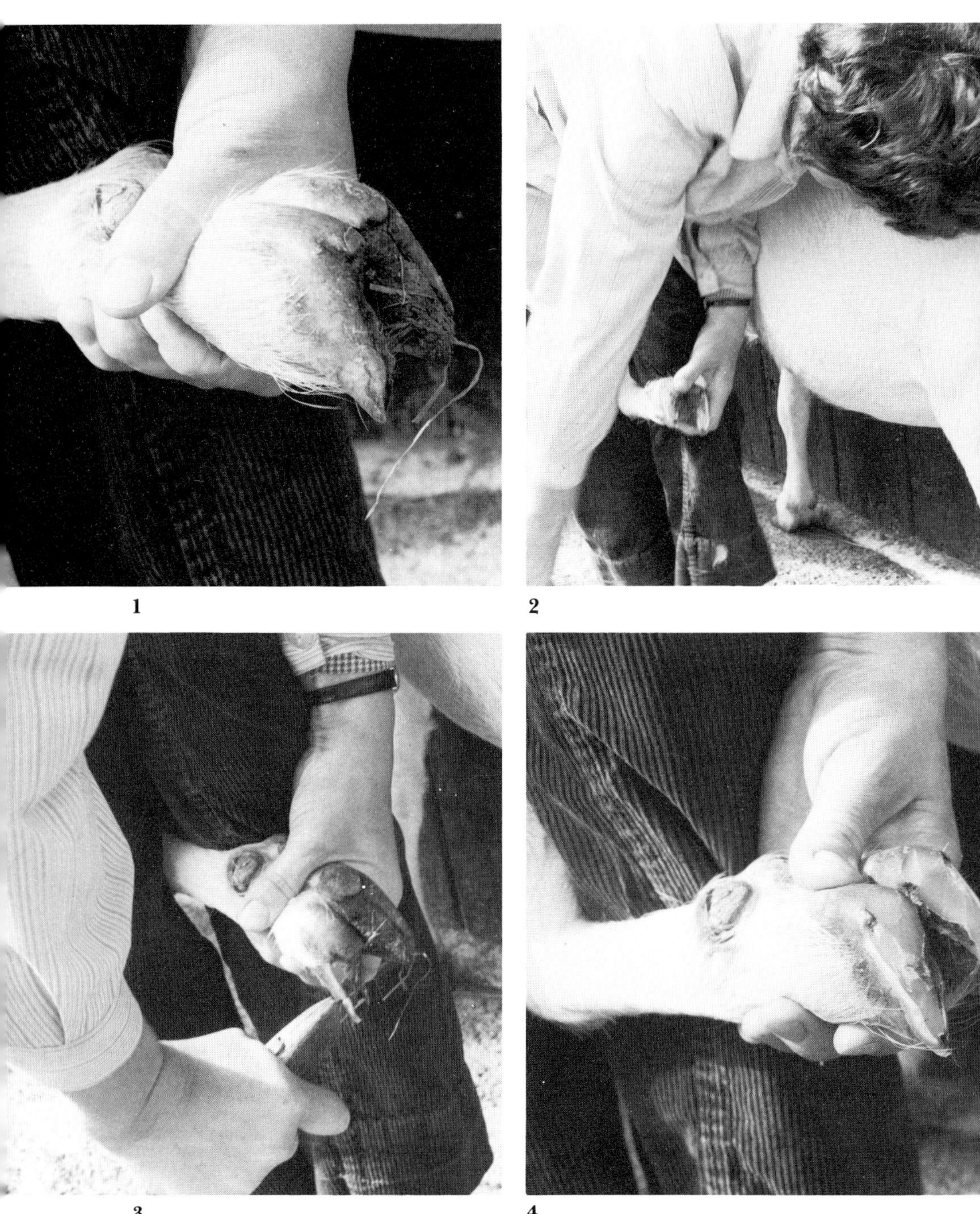

1

2

3

4

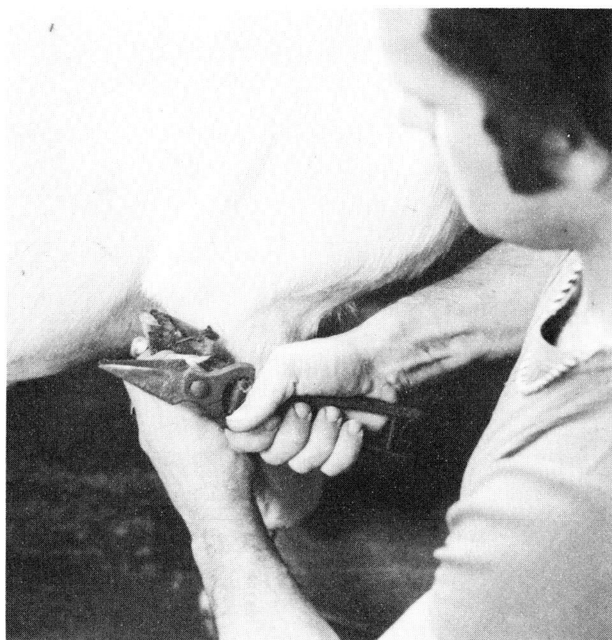

Hooftrimming
1. A hoof in need of trimming. **2.** The foreleg is held firmly between the herdsman's knees. **3.** The outer, overgrown horny wall of the hoof is cut away. **4.** The trimmed hoof.

Right: Using foot rot shears to trim the hoof.

tle (preferably not glass) or with a disposable plastic syringe. The goat's mouth is opened by the attendant putting the palm of the hand on the bridge of the goat's nose, and slipping fingers and thumb into the goat's mouth between the incisors and molars. As the goat opens its mouth, the bottle or syringe is slipped in towards the throat, and the liquid poured down. The goat's tongue should not be held down by the attendant, since the goat must be free to use it to aid swallowing.

Temperature taking

A clinical thermometer is shaken to ensure that the mercury column is below the scale and continuous. It is inserted gently, using a twisting movement, into the anus for a short way, and held there for one minute, before being slowly removed and read. It should then be cleaned and disinfected before being put away.

Normal rectal temperature	102.5-103°F = 39°C
Normal pulse rate (top of front leg, inner side)	70-95/minute
Normal respiration rate	20-24/minute

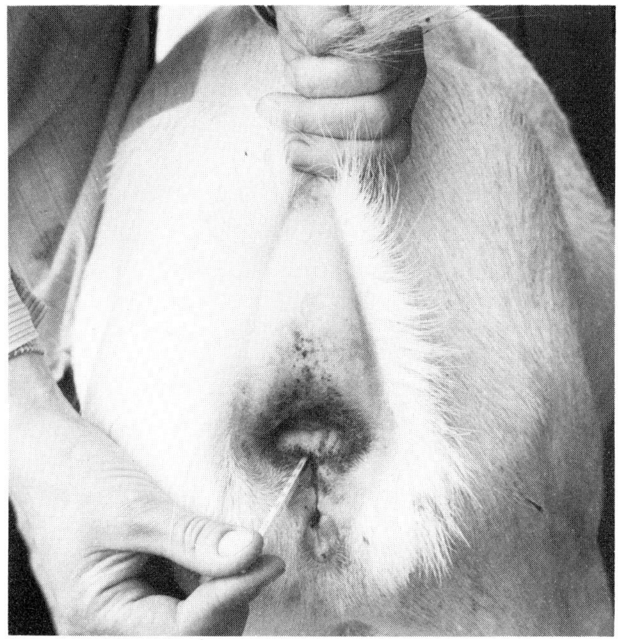

Taking a goat's temperature

Disposal of cull and elderly stock

Unwanted kids and any that are congenitally deformed should be killed as soon as possible after birth. (Although some people will buy in kids for meat, this should be arranged beforehand, so that any born but not required can immediately be disposed of). Veterinary surgeons will usually put kids down, or alternatively they can be killed at home using chloroform. A wide necked jar (e.g., sweet jar) is laid on its side, a wad of cotton wool dampened with chloroform put in the bottom and the kids nose placed in the jar. It must not be jammed in so that air cannot enter. After a couple of minutes, the kid will be deeply unconscious and the head is then pushed right in so that it breathes in chloroform only. Leave the kid for ten minutes or longer by which time it will be dead.

An alternative method is to bang the kid on the back of the head with a heavy implement, for example, a hammer, but it is recommended that anyone attempting this should first watch someone skilled doing it to prevent any risk of error. Drowning should not be attempted as it is a prolonged, unpleasant death for the kid. Kid carcasses can be buried, or can be used for meat as long as chloroform is not used.

Older stock present more of a problem for disposal. Unless the goat owner has ideological objections, hunt kennels will provide the service of coming to shoot elderly or infertile cull stock, and removing the carcass. Knackers provide a similar service, and will remove carcasses of animals that have died of an illness.

Nursing the sick goat

Whilst severe illness is not common in goats, it does occur from time to time and goat-keepers should be prepared to cope with a sick goat. A sick ruminant has the particular problem that, if the rumen ceases to function, a build-up of gas occurs which makes the animal 'blown'. A rumen that ceases to function is often difficult to get to work again. A goat that is too ill to stand should not be allowed to lie on its side, but should be propped up between straw bales. Every attempt should be made to get it to move, and preferably stand, as frequently as possible. Warmth is essential for a sick goat. A warm rug should be put on the goat or hessian sacking or a piece of blanket tied round it. The animal must not lie in a draught.

Warm salted water with molasses or black treacle added to it should encourage the goat to drink, and titbits to tempt it to eat should be offered. Ivy twigs (without berries), raspberry and blackberry shoots are excellent appetizers, and soft sweet hay can be offered from time to time. When once the goat is beginning to recover, more greens and cut grass and herbs can be offered. Bran is excellent to feed before re-introducing the usual concentrate mix.

First aid

First aid is the help that the goat owner can give on first seeing that a goat is ill or injured. Prompt correct treatment can often prevent a condition deteriorating before veterinary assistance arrives. A good stockman will always be casting an eye over his stock, and will know when an animal is off-colour or is behaving unusually.

It is advisable to keep a basic first aid kit at hand for treating accidents and the first signs of disease.

First aid box—suggested contents

Antiseptic lotion, e.g., Dettol, Savlon	Drenching bottle
	Clinical thermometer
Antiseptic cream	Udder cream
Cotton wool	Kaolin powder
Surgical gauze	Liquid paraffin
Gauze bandage (5 cm [2 in])	Epsom salts
Crepe bandage (5 cm [2 in])	Worming formula
Scissors (rounded ends)	Calcium borogluconate

Diseases and illnesses of goats

The following list is of those illnesses to which goats are most prone, or which, if they do occur, are most serious.

Abscesses

Abscesses appear as lumps under the skin, and are treated by the veterinary surgeon (or skilled goat-keeper) by lancing, cleaning out with antiseptic solution, and keeping the wound open for several days with regular applications of antibiotic cream. In the United States, a much more severe condition occurs called caseous lymphadenitis, which always requires veterinary treatment and can be fatal.

Accidents

Cuts and wounds should be treated by bathing with antiseptic solution. If at all severe, veterinary assistance should be called since suturing may be required, and antibiotic and anti-tetanus cover will almost certainly be needed. Fast-flowing blood should be checked with direct pressure using pads of gauze or cotton material, never by tourniquet. Wounds to the udder should always be considered as serious, since udder tissue easily becomes infected.

Fractured limbs, particularly in kids, can occur, and require setting. Keep the animal as still as possible until veterinary help arrives.

Acetonaemia

Acetonaemia occurs just before, or shortly after, kidding when a goat's energy requirements are particularly high. Sudden loss of condition and appetite, and drop in milk yield occur. The breath of the goat may or may not smell of 'pear drops' or nail varnish remover. In severe cases the animal goes into a coma. A goat which is fat and eating large quantities of concentrates is particularly prone to acetonaemia.

Treatment is to drench the goat with 0·25 litre (8 oz) of propylene glycol, and repeat twice daily for two days. Alternatively, drench with 25 g (1 oz) of bicarbonate of soda in half a cup of water, followed by 0·25 litre (8 oz) honey. A goat with acetonaemia requires veterinary attention, and the vet will advise on further treatment.

Anthrax

Anthrax is a rare, highly infectious disease. It is nearly always fatal and is rarely preceded by any noticeable symptoms. It is a notifiable disease (see Chapter 12).

Bloat (blown, hoven)

When the rumen ceases to function properly for some reason, a build up of gas occurs causing bloat. The goat's abdomen will appear distended, particularly on her left flank, and she will be in considerable pain. Liquid (medicinal) paraffin (20 to 50 ml depending on the size of the goat) is administered as a drench, and every attempt is made to keep the goat on her feet and moving. If no improvement occurs within an hour, call veterinary assistance.

The condition can be brought on by feeding too much green food, particularly wet grass and legumes or frosted greens. A good precaution is to feed hay before feeding goats greenstuff.

Brucellosis See Medical Aspects in Chapter 9.

Caprine Arthritis Encephalitis (CAE)

Although the incidence of infection is significant, the disease is rare showing as arthritis in older goats. Both encephalitis of kids and arthritis are more common in the USA and Australia.

Coccidiosis

This is most frequently seen in kids, causing scouring and poor condition. Treatment is with sulphamethazine, which is prescribed by a veterinary surgeon. Older stock may also be affected, and the condition tends to be more persistent in adults. The symptoms are general unthriftiness and poor condition with scouring or constipation, often occurring alternately. Goats cannot be infected with coccidia from rabbits or poultry, but can be infected by coccidia carried by sheep.

Colic

Colic is a general symptom of digestive upset arising from various causes (e.g., intestinal blockage, poisoning). It results in severe pain, but without the ruminal distention characteristic of bloat. Administration of liquid paraffin (50 ml) or Epsom salts (2 tablespoons in water) can be tried. If the condition does not ease, seek veterinary advice.

Diarrhoea (scours)

Scouring in milk-fed kids is not unusual, and replacing a feed with warm water, glucose and a dessertspoon of kaolin powder should clear the condition. If it does not, coccidiosis or worms should be suspected. In adults, suspect worms and treat accordingly, but if the condition does not improve within hours, veterinary advice should be sought before the animal becomes dehydrated. Administration of a hypotonic electrolyte solution will help to prevent dehydration. To 1 litre (2 pt) warm water add ½ teaspoon table salt, ¼ teaspoon bicarbonate of soda and 2 tablespoons dextrose or honey.

Administer by drenching at the rate of 0.5 litre (¾ pt) per 4.5 kg (10 lb) bodyweight per day.

Enterotoxaemia

Prevention of clostridial diseases by vaccination has already been described.

An affected goat suddenly loses its milk, ceases to eat, is in considerable pain and stretches its abdomen, and becomes unable to balance. Death follows rapidly. Immediate veterinary treatment may save a goat, but the disease is frequently fatal. It strikes particularly at the healthy, thriving animal and can be precipitated by sudden change in diet, gorging concentrates or lush pasture, or travelling, particularly after a concentrate feed. Vaccination is strongly recommended.

Foot and mouth disease

Occasional outbreaks occur in Britain, and the disease spreads rapidly, affecting all cloven-hoofed animals. As with anthrax, the disease is notifiable. Symptoms are blisters inside the mouth and around the top of the hooves.

Foot rot

Foot rot is highly infectious and causes severe lameness in sheep and goats. The hoof is killed by the invading organism, and a characteristically foul smell occurs. Because goats' feet receive more frequent attention than sheeps', it is not usually as much of a problem as in sheep flocks. The condition is treated by cutting away the dead part of the hoof, dipping the foot in 10% formalin solution, and spraying with Terramycin foot rot spray. The feet should be checked regularly until the condition is clear, and the animals moved onto clean pasture.

Goat pox

True goat pox is unusual in Britain and is characterized by spots on bare skin, particularly the udder, and a fever. Many 'spotty' conditions of the udder are loosely referred to as goat pox, but treatment for all of them is similar. Wash the udder with mild antiseptic solution, dry thoroughly and dust with boracic powder. Do not use cream on the udder.

Laminitis

This is an inflammation of the soft part of the hoof, usually occurring shortly after kidding, or after feeding too high a level of protein. It causes considerable pain and lameness. The condition can be eased by bathing the foot with warm water, then applying a bran poultice. If the condition persists, a vet can administer drugs to reduce the swelling.

Mastitis

Mastitis is an infection of the udder tissue, and can be caused by a variety of factors, such as infection carried by flies from another goat or on the hands of the person milking, or from physical damage, such as a bang or cut or rough handling during milking.

In mild cases, the milk may taste 'salty' and the milk, particularly the foremilk, will contain string-like clots. Lumps may well be felt in the udder. If treated rapidly with antibiotics, cases of this kind can be cured relatively easily.

The acute forms are more severe. Within a matter of hours the goat has a high temperature and the udder becomes hot, swollen and hard. Immediate action is necessary to save the goat, and even so, part of the udder may remain permanently damaged. The vet will give antibiotic treatment, and the udder should be bathed alternately with hot and cold water, and milked as frequently as possible to alleviate pressure on the udder tissue. The goat will be ill, and will require careful nursing to get her to eat and to begin to recover.

The milk from a goat with any form of mastitis must be discarded until after the antibiotic treatment is finished. The goat should always be milked last until the infection is cleared.

Metritis

This is an infection of the uterus, occuring most frequently within a few days of kidding. A foul-smelling discharge comes from the vulva, the animal runs a temperature and is generally unwell. Antibiotics, usually in the form of pessaries, will cure the condition. Any manual interference during kidding should be followed by antibiotic cover as a preventive measure.

Milk fever

Milk fever is, strictly speaking, a misnomer since body temperature is not raised, and is frequently depressed. It is caused by a sudden depletion of calcium from the blood stream. It is commonest at the start of lactation, in older animals and in heavy producers, particularly those fed large amounts of concentrates to force milk production soon after kidding. Typically, the goat becomes unsteady on her feet and listless. In some cases the goat may collapse, although this is not as common as it is with cows. Calcium borogluconate (50 ml) injected subcutaneously gives rapid recovery.

Navel ill

In this condition, the navel of a very young kid becomes swollen and painful to touch, and the kid is frequently unwell. Antibiotic treatment is required since the condition can progress to generalized septicaemia or inflamed joints, leaving the kid with an arthritic condition. Navels should always be dipped in tincture of iodine or sprayed with a Terramycin aerosol spray as soon as possible after birth as a preventive measure.

Poisoning

The following plants are poisonous, and care should be taken to ensure that they do not grow on land to which goats have access:

```
   Arum, Cuckoo pint (Arum maculatum)
  *Autumn crocus (Colchicum autumnale)
   Bracken (Pteridium aquilinum)
   Bryony—black (Tamus communis)
           —white (Bryonia dioica)
   Dogs mercury (Mercuralis perennis)
   Foxglove (Digitalis purpurea)
  *Hemlock (Conium maculatum)
   Horsetails (Equisetum spp) — particularly in hay
   Ivy (Hedera helix) — berries only
   Laburnum (Laburnum anagyroides)
   Laurel (Prunus laurocerasus)
   Nightshades (Atropa and Solanum spp)
   Oak (Quercus spp) — acorns
   Potato (Solanum tuberosum) — haulm and green tubers
   Privet (Ligustrum vulgare)
   Prunus spp — withered leaves of plum, cherry etc.
  *Ragwort (Senecio jacobaea) — particularly in hay
  *Rhododendron (Rhododendron ponticum)
  *Rhubarb (Rheum raponticum) — leaves
   Spindle tree (Euonymus europaeus)
  *Thorn-apple (Datura stramonium)
   Tomato (Lycopersicum esculentum) — foliage
  *Water dropwort (Oenanthe crocata)
  *Yew (Taxus bacata)
```

Plants marked with an asterisk are particularly poisonous.

If a goat is known to have eaten any of them, veternary advice should be sought immediately.

The book *Poisonous Plants in Britain* (Bulletin 161), published by the Ministry of Agriculture, Fisheries and Food, is invaluable for reference. In addition to describing the plants and their effects it also details methods of first aid treatment.

Pneumonia

Pneumonia is commonest where housing is poorly ventilated. If condensation forms inside the house, ventilation is inadequate and the goats are at high risk. Pneumonia can also be a complication following another disease. Antibiotic treatment should be given as soon as possible after the animal becomes unwell. Coughing and audible breathing are evident when the illness is established. Mortality is frequently high.

Pregnancy toxaemia

Pregnancy toxaemia is a similar condition to acetonaemia, generally occuring during the last two months of pregnancy. A goat fed particularly well in the early stages of pregnancy is most susceptible. The developing kids begin to rob the doe of her body reserves, and she becomes unwell and unable to eat. The condition is dangerous to kids and doe, so veterinary assistance should be sought.

Rickets

Rickets occurs in kids unable to get sufficient Vitamin D and calcium for healthy bone growth. Winter kids or those born in a year with little sunshine are particularly susceptible. The first symptoms are swollen joints, followed by a bent appearance of the long bones. Vitamin D preparations can be prescribed by the vet, and sterilized bone meal sprinkled on the goat's food will help to halt the condition.

Skin diseases

Eczema

Eczema is a scabby condition, not infectious, which affects some animals, particularly kids and goatlings, often during the winter months. The scabs are particularly persistent round the eyes and mouth. Antibiotic injections, change of diet (e.g., removing maize and sugar beet pulp and increasing greenfoods), and treatment with various skin preparations may ease the condition. It is unsightly, but not serious except that the animal is at risk to secondary infections. Frequently the condition clears after first kidding and does not recur.

Lice

Lice are blood-sucking skin parasites, causing irritation and hair loss. Regular dusting with a proprietary louse powder should be part of the husbandry programme.

Mange

Mange is another parasitic condition, caused by mites which burrow into the skin resulting in hair loss, scabbing and irritation to the animal. Bathing with BHC or Coumaphos preparations may clear the condition, but if it persists, alternative preparations may be provided by the veterinary surgeon. It is infectious, though some goats are more susceptible to infestation than others.

Orf (contagious pustular dermatitis)

Orf is a viral condition, common in sheep. It causes pustules around the mouths of the animals and if kids are suckling, udders can become infected too. Antibiotic aerosol treatment will aid healing. After infection, a goat is likely to be immune to the condition for approximately a year. If orf is a particular local problem, a vaccination programme should be considered. The disease is transmissible to man and protective clothing (including rubber or disposable surgical gloves) should be worn when handling infected animals.

Tetanus

Tetanus, like enterotoxaemia, is a clostridial disease, which can be prevented by vaccination. It is frequently fatal. The symptoms are general muscular stiffness, particularly of the neck and jaw.

Worms (internal parasites)

All goats carry a small roundworm burden, but if conditions are such that the number of worms in the gut builds up rapidly, the goat can lose condition. High risk times are immediately after kidding and in the late spring when the grass is lush and the weather showery. Kids under six months old put out to graze on infected pastures are particularly susceptible. Worming programmes have been discussed earlier in this chapter.

Goats occasionally get tape worms and these can be diagnosed by the veterinary surgeon, who will prescribe a suitable drug as a vermifuge.

Liver flukes can also occur in goats. For goats grazing on low wet pastures, where cattle and sheep are known to be at risk, it is wise to use a worming preparation which includes an anti-fluke compound. Fluke infestation can cause permanent damage to the liver, so preventive measures are strongly advised.

12 Goats and the law

Whilst the Ministry of Agriculture involves itself as little as possible with goats, various laws applicable to livestock generally, and to food production, do affect the goat-keeper. Several European countries have sizeable goat industries, and are consequently more affected by legal matters, and EEC legislation may well come to embrace goats to a greater extent than we in Britain have ever been used to.

Laws change or become amended from time to time, so if a situation arises where a goat-keeping enterprise comes up against the law, it is as well to seek professional advice.

Straying

It is the responsibility of the owner to fence in his own stock securely. Even if a trespasser leaves gates open or breaks fences, and animals escape, the owner can be held liable for any damage caused, whether to growing crops, private property or in a road accident. For this reason, it is advisable to take out third party insurance to cover straying.

Notifiable diseases

Several diseases, for economic or public health reasons, are notifiable by law. Any livestock owner who suspects that an animal he owns is suffering from one of these diseases is required to notify the authorities. In practice, a veterinary surgeon would normally be contacted and would alert the police and MAFF if he diagnosed a notifiable disease. Restrictions on movement or even compulsory slaughter will be imposed. Anthrax and foot and mouth disease come into this category. All sudden unexplained deaths or illnesses amongst livestock should be diagnosed by a veterinary surgeon.

Movement records

Owners of livestock, including cattle, sheep, pigs and goats, are required by law to keep a record of movement onto and off their premises. This must include destination and origin and the date of the journey. The Consumer Protection Department of the County Council is empowered to inspect these at any time, but often they will be checked only in the event of an outbreak of an infectious disease in the area. Printed books can be bought for the purpose, or an exercise book can be ruled as below:

Date	Number of animals	Premises from which moved	Premises to which moved
20.3.80	1 goat	Hill Farm, Sutton	Veterinary Surgeon, Low Road, Sutton and return

Unlike cattle, goats, sheep and pigs do not have to be individually identified.

Transport of goats

Goats being transported for the purpose of breeding, exhibition or as pet animals, which have been kept together on the same premises, may be carried together so long as goats over six months of age are separated from younger goats.

Goats being transported for purposes other than breeding, exhibition or as pets must be separated into four categories: female goats with accompanying unweaned kids, male goats over six months of age, weaned kids under three months of age and other goats. Purpose built vehicles with loading ramps are not essential.

Anaesthesia

The only procedure that requires anaesthesia that a goat-keeper is likely to perform on his goats is castration. Under a week of age, castration can be carried out without anaesthesia.

Sale of milk products

The legal situation concerning sales of goats milk and milk products is described in Chapter 9.

13 Selling stock and exhibiting

Advertising

Some form of advertising is usually necessary to sell stock, though at times goats, particularly milkers, are in such demand, that merely to mention a goat is for sale may well bring suitable customers.

Advertisements can be placed in various goat societies' publications, such as the British Goat Society's Monthly Journal, and newsletters of its Affiliated Societies. Advertisement rates are usually not excessive. If the stock is well bred it should sell easily through these media.

Local and regional newspapers accepting classified advertisements are another useful source, but more care is required in ensuring that the home the goat is going to is suitable. Dealers make frequent use of these columns, as, in some areas, do certain ethnic groups requiring goats for ritual slaughter.

The secretaries of some of the Affiliated Societies keep lists of stock for sale and purchasers' requirements. It is worth putting goats onto these lists in the hope of placing them.

Agreements

As with any sale, the vendor is required by law to act honestly. It is unfair to the buyer, particularly a beginner, to fantasize on the goat's achievements and merits. It is only good salesmanship to describe the animal in the best terms possible, as long as any particular drawbacks or faults are pointed out.

Any history of illness, such as mastitis or acetonaemia, which may well recur, should be mentioned, as should age, number of kiddings and points such as whether the goat will tether successfully or not. The vendor may wish to impose terms on the sale, such as first refusal of the animal should the buyer wish to sell it again at a later date. Some in-kid goats are sold on breeding terms where the vendor takes a particular kid and does not transfer the goat to the new owner until after the kidding, ensuring that kids are registered with the vendor's prefix.

Should a goat, once sold, not settle, or develop a disease that can be traced to its previous home, or prove infertile, it is usual practice for the vendor to either take the goat back or come to some mutually agreeable financial arrangement with the new owner.

Forms of agreement for a sale are obtainable from the British Goat Society and help to make the sale agreement legally more watertight. All registered stock must be transferred to its new owner by the previous owner. A transfer of ownership form (obtainable from the British Goat Society) is filled in and sent to the Society Secretary with the registration card. The Secretary then, having recorded the transfer, sends the card on to the new owner.

Exports

A small steady export trade of goats from this country has existed for many years, but, of late, it has increased considerably in size. Goats go to many parts of the world, particularly Europe, Africa, South America and the Far East. Some countries (e.g., Australia) will not accept any imports from Britain, and others (e.g., America) make it practically impossible.

The negotiations and paperwork involved in arranging an export are extremely complex, and it is always wisest to do any exporting through an agent. The prices obtained for exported stock can be considerably higher than those obtained in this country, but exporting can involve months of waiting, which, in the end, makes an early home sale more profitable. All stock for export must be of good breed type, and in sound health, preferably from lines of proven ability. Extensive veterinary tests, and frequently quarantine periods, are involved. The standard of British stock in general may be judged from a few exported animals.

Exhibiting

Shows are the shop window of the goat world. Many are held in all parts of the country during the summer months. Some are associated with County agricultural shows, others are organized by local clubs or individuals. The British Goat Society gives official recognition to the major shows, particularly those with milking competitions or classes for male goats. Many shows are unrecognized, but nevertheless, the standard of both entries and organization remains high.

The recognized shows are listed in the British Goat Society Journal in the January/February and March editions, and local unrecognized shows are advertised in club magazines. Probably the most important part of all showing is the milking competition.

Further information

Societies and clubs

British Goat Society
The British Goat Society, founded in 1879, is the premier goatkeepers' society in Great Britain. Amongst its activities it publishes an annual *Herd Book* giving details of stock registered with the Society, a *Year Book* consisting principally of articles on aspects of goats and goat-keeping, and a *Monthly Journal*. It also produces a number of useful booklets and leaflets. Leading shows throughout the country are recognized by the BGS.

Mrs S. May, Secretary,
British Goat Society,
Rougham,
Bury St Edmunds,
Suffolk, IP30 9LJ

Beyton (03597) 0351

A stamped self-addressed envelope will be appreciated with enquiries.

Local goat-keeping organizations
There are now numerous local clubs and societies, organized on a regional, county or district basis throughout Great Britain and affiliated to the British Goat Society. Such clubs arrange educational and social meetings, organize shows and, in many cases, publish newsletters and magazines for their members.

Officials and members of local societies are excellent sources of information, help and advice, particularly for beginners, regarding things such as the local availability of stock, stud males, feeding stuffs, materials and equipment. Secretaries of local societies change from time to time. Rather than give details of societies here, we suggest that those interested write to the British Goat Society for an up-to-date list.

Goat Veterinary Society
Although membership of this society (formed in 1979) is restricted principally to veterinarians, a Journal containing scientific papers is published and can be obtained by non-members. Further details are available from:

John G. Matthews, BSc, BVMS, MRCVS
The Limes,
Chalk Street,
Rettendon Common,
Chelmsford,
Essex.

Goat Producers Association
Mrs J. Barley, FIAT
c/o AGRI
Church Lane,
Shinfield,
Reading,
Berks.

Publications—General
Goats, H. E. Jeffery (CASSELL, 1975)
Goat Husbandry, D. Mackenzie (FABER, 1980)

Goat Production in the Tropics, C. Devendra and M. Burns (COMMONWEALTH AGRICULTURAL BUREAUX, 1970)
Exhibition and Practical Goatkeeping, J. Shields (SPUR PUBLICATIONS, 1977)
Management and Diseases of Dairy Goats, S. Guss (DAIRY GOAT JOURNAL CORPORATION, 1977)
Home Goat Keeping, L. Hetherington (EP PUBLISHING 1977)
Goatkeeping, (British Goat Society, [BGS] 1976)
Breeds of Goats, (BGS, n.d.)
Life Story of a Goat (BGS)
Dairy Goatkeeping (AL 118), (MINISTRY OF AGRICULTURE, FISHERIES AND FOOD [MAFF])
Commercial Goat Farming, K. Thear (BROAD LEYS PUBLISHING CO 1985)

Dairying

Making Cheeses, S. Ogilvy (BATSFORD, 1976)
Home-made Butter, Cheese and Yoghurt, M. Black (EP PUBLISHING, 1977)
The Home Dairying Book ed. K. Thear (BROAD LEYS PUBLISHING COMPANY, 1978)
Dairy Work for Goatkeepers, (BGS, 1973)
Cream, (AL 495), (MAFF)
Clotted Cream, (AL 438), (MAFF)
Cream Cheese, (AL 222), (MAFF)
Taints in Milk, (AL 322), (MAFF)
Soft Cheese, (AL 458), (MAFF)
Hand Cleaning of Dairy Equipment, (AL 422), (MAFF)
Starters for Cheesemaking, (AL 302), (MAFF)

Feeding and management

Goat Feeding, (BGS, n.d.)
Wild Food for Goats, (BGS)
Electric Fencing, (Bulletin 147), (HMSO, 1969)
Poisonous Plants in Britain (Reference Book 161), (HMSO, 1984)

Rations for Livestock, (Bulletin 48), (HMSO, 1970)
Sugar Beet Pulp for Feeding, (AL 363), (MAFF)
Growing Grass for Hay, (AL 395), (MAFF)
Hay: Quality for Feeding, (SL 488), (MAFF)
Lucerne, (AL 67), (MAFF)
Feeding Separated Milk, (AL 439), (MAFF)

Serial publications

British Goat Society Journal, BGS (11 issues per year)
British Goat Society Year Book, BGS (annual)
British Goat Society Herd Book, BGS (annual)
Home Farm BROAD LEYS PUBLISHING Co., (bimonthly)

Suppliers

Below is a list of suppliers who are prepared to send items by post. Many areas have local suppliers and farmers co-operatives, which can be sources of materials and equipment.

Astell Lab. Services Co.,
172, Brownhill Road,
Catford,
London, SE6 2DL.
Dairy thermometers

Blow, J. J. Ltd.,
Oldfield Works,
Chatsworth Road,
Chesterfield,
Derbys, S40 2DJ.
Milking pails, strainers, churns etc.

Boddington, W. H. & Co. Ltd.,
Horsmonden,
Tonbridge,
Kent, TN12 3AH.
Cheese moulds

Bowater Industrial Packaging Ltd.,
Princes Way,
Team Valley Estate,
Gateshead, NE11 0UT.
Cartons for milk, yoghurt and cheese

British Goat Society,
Rougham,
Bury St Edmunds,
Suffolk, IP30 9LJ.
Books, leaflets, forms etc.

Dellrose Dairy Products,
The Gate House,
Millstone Lane,
Oakerthorpe,
Derbys.
Dairy equipment, milking equipment etc.

Fullwood & Bland Ltd.,
Ellesmere,
Salop, SP12 9OG.
Rennet, annatto etc.

Gascoigne Milking Equipment Ltd.,
Berkeley Avenue,
Reading,
Berks, RG1 6JW.
Milking machines

Government Bookshop (HMSO),
P.O. Box 569,
London, SE1 9NH.
HMSO books and publications

Hansen, Chr., Laboratories Ltd.,
476, Basingstoke Road,
Reading, RG2 0QL.
Cultures for yoghurt, cheese etc.

Harvester Goat Shop,
Maylord Street,
Hereford.
Goatkeeping equipment

Horstmann Gear Co. Ltd.,
The Newbridge Works,
Bath, BA1 3EF.
Earmarking equipment

Hunters of Chester,
Chester.
Grass and clover seeds, herbal mixtures etc.

Lakeland Plastics Ltd.,
Alexandra Road,
Windermere.
Plastic milk bags, sealers

Landsmans Bookshop Ltd.,
Buckenhill,
Bromyard,
Herefordshire.
Books

Ministry of Agriculture, Fisheries &
Food
Lion House,
Alnwick,
Northumberland.
Advisory leaflets (mostly free of charge)

Nutrichip Products,
21 Gastard,
Corsham, Wilts.
Minerals

Ritson, F.,
Goat Appliance Works,
Longtown,
Carlisle.
Goatkeeping equipment

Self Sufficiency & Smallholding
Supplies,
The Old Palace,
Priory Road,
Wells,
Somerset, BA5 1SY.
Goatkeeping and smallholding equipment

Small Scale Supplies,
Widdington,
Saffron Walden,
Essex, CB11 3SP.
Books

Watkins & Doncaster,
Four Throws,
Hawkhurst,
Kent.
Lankroline F.P.4 for curing skins

Index

Page numbers in italic refer to illustrations